More information about this series at http://www.springer.com/series/4748

Ganesh R. Naik
Editor

Non-negative Matrix Factorization Techniques

Advances in Theory and Applications

Editor
Ganesh R. Naik
University of Technology Sydney
Sydney, NSW
Australia

ISSN 1860-4862 ISSN 1860-4870 (electronic)
Signals and Communication Technology
ISBN 978-3-662-48330-5 ISBN 978-3-662-48331-2 (eBook)
DOI 10.1007/978-3-662-48331-2

Library of Congress Control Number: 2015948772

Springer Heidelberg New York Dordrecht London

Printed on acid-free paper

Springer-Verlag GmbH Berlin Heidelberg is part of Springer Science+Business Media
(www.springer.com)

Preface

Nonnegative Matrix Factorization (NMF) is one of the widely used matrix factorization techniques for revealing hidden factors that underlie sets of random variables, measurements, or signals. NMF is essentially a method for extracting individual signals from mixtures of signals. The concept of a matrix factorization arises in a variety of significant applications and each matrix factorization task makes a different assumption regarding factor matrices and their underlying configurations. Therefore choosing the appropriate one is important in each application domain. In most instances, the data to be analyzed are nonnegative, and sometimes they also have sparse representations. For such applications, it is preferable to take these constraints into account in the analysis to extract nonnegative and sparse/smooth components or factors with physical meaning or reasonable interpretation, and thereby avoid absurd or unpredictable results. NMF research can be motivated by the open problems and continuing research on these problems, and hence a need to edit this book to report the latest results on the topic. These challenges motivate further research in the area of NMF, and this book intends to put together all the innovative ideas and new results in this research area.

This book aims to disseminate timely to the scientific community the new developments in NMF spanning from theoretical frameworks, algorithmic developments, to a variety of applications. I believe that a potential impact of the NMF and its extensions on scientific advancements might be as great as the other popular matrix factorization techniques such as Independent Component Analysis (ICA) or the Singular Value Decomposition (SVD) and Principal Component Analysis (PCA). In contrast to ICA or SVD/PCA approaches, NMF technique, if successively realized may improve interpretability and visualization of large-scale data while maintaining the physical feasibility more closely.

The book provides a wide coverage of the models and algorithms for NMF both from a theoretical and a practical point of view. It also covers some emerging techniques in NMF, especially those developed recently, offering academic researchers and practitioners a comprehensive update about the new development in this field. The book provides a forum for researchers to exchange their ideas and to foster a better understanding of the state of the art of the subject. I envisage that the publication of this book will stimulate new ideas and more cutting-edge research activities in this area.

This book is intended for computer science and electronics engineers (researchers and graduate students) who wish to get novel research ideas and some training in NMF, sparse component analysis, machine learning, artificial intelligence, and signal processing applications. Furthermore, the research results previously scattered in many scientific articles worldwide are methodically collected and presented in the book in a unified form. As a result of its twofold character the book is likely to be of interest to researchers, engineers, and graduates who wish to learn the core principles, methods, algorithms, and applications of NMF.

I would like to thank the authors for their excellent submissions (chapters) to this book, and their significant contributions to the review process which have helped to ensure high quality of this publication. Without their contributions, it would have not been possible for the book to come successfully into existence. In addition, my special thanks to Ms. Lu Yang and Ms. Jessie Guo from Springer Beijing for their kind support and great efforts in bringing the book to fruition.

Sydney
June 2015

Ganesh R. Naik

Contents

Chapter 1
From Binary NMF to Variational Bayes NMF: A Probabilistic Approach

R. Schachtner, G. Pöppel, A.M. Tomé and E.W. Lang

Abstract A survey of our recent work on probabilistic NMF is provided. All variants discussed here are illustrated by their application to the analysis of failure patterns emerging from manufacturing and processing silicon wafers. It starts with *binNMF*, a variant developed to apply NMF to binary data sets. The latter are modeled as a probabilistic superposition of a finite number of intrinsic continuous-valued failure patterns characteristic for the manufacturing process. We further discuss related theoretical work on a semi-non-negative matrix factorization based on the logistic function, which we called *logistic NMF*. While addressing uniqueness issues, we propose a *Bayesian Optimality Criterion* for NMF and a determinant criterion to geometrically constrain the solutions of NMF problems, leading to *detNMF*. This approach also provides an intuitive explanation for the often used multilayer approach. Finally, we present a *Variational Bayes NMF* (VBNMF) algorithm which represents a generalization of the famous Lee–Seung method. We also demonstrate its ability to estimate the intrinsic dimension (model order) of the NMF method.

R. Schachtner · E.W. Lang (✉)
CIML/Biophysics, University of Regensburg, 93040 Regensburg, Germany
e-mail: elmar.lang@ur.de

R. Schachtner · G. Pöppel
Infineon AG, 93040 Regensburg, Germany
e-mail: reinhard.schachtner@infineon.com

G. Pöppel
e-mail: gerhard.poeppel@infineon.com

A.M. Tomé
DETI/IEETA, Universidade de Aveiro, 3810-193 Aveiro, Portugal
e-mail: ana@ieeta.pt

G.R. Naik (ed.), *Non-negative Matrix Factorization Techniques*,
Signals and Communication Technology, DOI 10.1007/978-3-662-48331-2_1

1.1 Motivation

Non-negative Matrix Factorization (NMF) is a blind signal decomposition technique approximating multivariate non-negative data sets by a linear superposition of underlying non-negative basis components and non-negative weights according to the model

$$\mathbf{X}^T = \mathbf{H}^T\mathbf{W}^T + \mathbf{E}^T \tag{1.1}$$

$$\mathbf{X}_{*n} = \sum_k \mathbf{W}_{*k} H_{kn} + \mathbf{E}_{*n} \tag{1.2}$$

where the $M \times N$-dimensional matrix $\mathbf{X}^T$ represents a two-dimensional data array, the $M \times K$-dimensional matrix $\mathbf{H}^T$ represents the matrix of hidden factors, the $K \times N$-dimensional matrix $\mathbf{W}^T$ their corresponding weights in the superposition approximation and, finally, the $M \times N$-dimensional matrix $\mathbf{E}^T$ denotes the reconstruction error. Here N denotes the number of observation vectors, M the number of components of these vectors, and K a yet to be determined intrinsic dimension. Note that in this interpretation each column vector $\mathbf{X}_{*n}$ is represented in a K-dimensional basis system spanned by the K basis vectors $\mathbf{W}_{*k}$. To be more specific, in this contribution we will illustrate most of the discussed NMF variants by an application to so-called *wafer maps* which arise from manufacturing silicon wafers in a semiconductor factory. Each wafer encompasses a large number of semiconductor microchips and their production is controlled by several hundred functionality measurements. Only if a chip passes all measurements positively, it is labeled *pass*, otherwise it receives the label *fail*. In that way, a *pass/fail* pattern of every wafer is created which is called a *wafer map*. These characteristic patterns are now collected into an $N \times M$-dimensional data matrix $\mathbf{X}$. Thus the rows $\mathbf{X}_{n*}$ of the data matrix $\mathbf{X}$ represent N wafer maps reflecting observed *pass/fail* patterns of N wafers, each one containing M chips. Each such observed wafer map is then expanded into a system of K hidden *meta-wafer maps* $\mathbf{H}_{k*}, k = 1, \ldots, K$ and each row $\mathbf{W}_{n*}$ of the weight matrix $\mathbf{W}$ contains proper weights for the superposition of the basis vectors $\mathbf{H}_{k*}$ to approximate every observation $\mathbf{X}_{n*}$. These *meta-wafer maps* are supposed to reflect failure patterns characteristic of some deficiencies of the production process. To summarize we consider data decompositions of the form

$$\begin{aligned} \mathbf{X} &= \mathbf{WH} + \mathbf{E} \\ \mathbf{X}_{n*} &= \sum_k W_{nk}\mathbf{H}_{k*} + \mathbf{E}_{n*} \end{aligned} \tag{1.3}$$

This representation offers several aspects concerning the solution and applications of this matrix factorization problem:

- If the intrinsic dimension K is chosen according to $K \ll \min\{N, M\}$, a dimension reduction is achieved which represents the observations into a small number of characteristic *intrinsic maps*.

- Furthermore, the choice of reconstruction error **E**, for example Euclidean distance or one of the statistical divergences like Kullback–Leibler, Bregman, Itakura–Saito, Csiszar's, etc., determines the underlying error statistics and prescribes the route to solve the factorization problem with a machine learning technique. Note that by defining a suitable error function, the factorization approximation can be converted into an optimization problem.
- NMF has only rarely been applied to binary data, though in practice the latter occur frequently. We will deal with this problem and discuss the algorithms *binNMF* [1] and *logisticNMF* [2].
- The question of how to obtain a unique solution to the factorization problem still needs further research efforts to achieve a definitive answer. We introduce this problem with a geometric approach, while presenting our algorithm *detNMF* [3] and discussing its relation to a Bayesian NMF approach.
- Also model order selection, i.e., an optimal, problem-dependent choice of the inner dimension K needs further clarification. This problem is discussed in our algorithm *VBNMF* [4] which deals with a variational Bayes approach.
- Last but not least, efficient NMF algorithms for large-scale problems need to be developed [5].

The matrix factorization problem mentioned above falls into the category of blind source separation (BSS) problems. First attempts to solve the matrix factorization problem with added positivity constraints have been presented by Paatero and Tapper [6, 7]. Later, the seminal work of Lee and Seung in 1999 [8] boosted the field of non-negative matrix factorization. The authors discussed learning algorithms to optimize a cost function based on either a Euclidean distance measure or a generalized Kullback–Leibler divergence. The algorithms resulted in multiplicative update rules, thus yielding what might be called the canonical form of NMF update rules. These techniques are confined to deal with two-dimensional data arrays. However, recently the field expanded to multidimensional data arrays and is called Non-negative Tensor Factorization (NTF) [9].

In this contribution, we present a summary of our recent work on NMF extending from binary NMF to Variational Bayes NMF, and discuss illustrative applications mainly devoted to the analysis of failure patterns on silicon wafer maps.

First we deal with a binary abstraction of such failure patterns and present our work on *binary NMF* (binNMF) [1, 10]. It introduces a probabilistic variant of NMF applied to binary data sets. The latter are modeled as a probabilistic superposition of underlying continuous-valued elementary failure patterns. An extension of the well-known NMF procedure to binary-valued data sets is provided to solve the related optimization problem with non-negativity constraints. We further discuss related theoretical work on a semi-non-negative matrix factorization based on the logistic function, which we called *logisticNMF* [2]. This work presents a matrix factorization technique applied to the latent variables of the logistic function. The latter is used to model the occurrence probabilities of binary data sets. We provide simple learning and inference rules for this new model of binary data. The rules are derived by gradient ascent with a projection step forcing one of the factors to have only non-negative entries. *Logistic NMF* thus places itself between *probabilistic PCA* and *binary NMF*.

Next, we present a probabilistic treatment where we address uniqueness issues and propose a *Bayesian Optimality Criterion* (BOC) for NMF solutions, which can be derived in the absence of prior knowledge about the characteristics of the underlying factor matrices. Addressing uniqueness issues, we propose a determinant criterion to geometrically constrain the solutions of non-negative matrix factorization problems and achieve unique and optimal solutions in a general setting, provided an exact solution exists. We demonstrate how optimal solutions are obtained by a heuristic named *detNMF* [3, 11] in an illustrative example and discuss the difference to sparsity constraints. Furthermore, an intuitive explanation of the benefits of multilayer techniques, often applied in NMF algorithms, is discussed also. Finally, we present a new *Variational Bayes NMF* algorithm *VBNMF* [4, 12] which is a straightforward generalization of the canonical Lee–Seung method for the Euclidean NMF problem [13]. We also demonstrate its ability to automatically detect the actual number of components in non-negative data sets. Illustrative examples of the analysis of failure patterns on silicon wafer maps complete the theoretical deductions. We also show how the geometric *detNMF* approach, which we proposed earlier, naturally follows from such a Bayesian NMF approach.

1.2 NMF for Binary Data Sets

Binary data sets often provide a convenient abstraction of massive collections of measured physical quantities. An example is given by silicon wafer maps, which summarize the result of measuring hundreds of physical quantities on each single semiconductor chip on the wafer being manufactured by a simple *pass* or *fail* bit. Such fails can be induced in any of the numerous manufacturing steps, or be the result of a combination of some not perfectly calibrated or matched instances, or even be already present in the raw material. Obviously, failure events are of special interest and their binary abstractions often form characteristic patterns on the map. Such binary patterns can be understood as resulting from a superposition of elementary failure patterns which indicate sources of malfunction of specific equipment of the processing chain during wafer manufacturing. Considering a collection of N wafers, each containing M chips, all binary wafer maps will be arranged into an $N \times M$ data matrix $\mathbf{X}$.

1.2.1 Binary NMF: A Poisson Yield Model

The algorithm *binNMF* elaborates on this idea by modeling an observed failure pattern $\mathbf{X}_{n*}$ as a probabilistic superposition of K continuous-valued failure patterns $\mathbf{H}_{k*}, k = 1, \ldots, K$ and their related weights $W_{nk}, k = 1, \ldots, K$. The event *chip m on wafer n is pass* is encoded as $X_{nm} = 0$, while the event *chip m on wafer n is fail* is encoded as $X_{nm} = 1$.

Now we assume that a source k induces *fail* events on chip m of wafer n, which occur randomly and independently from each other. This allows us to apply a *Poisson yield model*, i.e., the *Poisson probability* of observing d fail events is given by

$$P_{\lambda_{nkm}}(d) = \frac{\lambda_{nkm}^d}{d!} \exp(-\lambda_{nkm}) \xrightarrow{d=0} \exp(-\lambda_{nkm}). \tag{1.4}$$

Let us now assume that each source k manifests itself by an intrinsic failure pattern $\mathbf{H}_{k*} = \{H_{k1}, \ldots, H_{kM}\} \geq 0$ across all chip positions. The impact of kth source S_k onto wafer n is reflected by the weight $W_{nk} \geq 0$. Thus the Poisson parameter can be modeled as $\lambda_{nkm} = W_{nk} H_{km} \geq 0$. It reflects the expected number of fail events induced by source k on chip m of wafer n, and where $P_{\lambda_{nkm}}(d = 0)$ indicates the probability that not even a single fail event has occurred.

Hence, we have for a single source S_k

$$P(X_{nm} = 0|S_k) = \exp(-W_{nk} H_{km}) \tag{1.5}$$

Given independent sources $S_k, k = 1, \ldots, K$ this yields the overall *pass* probability

$$\begin{aligned} P(X_{nm} = 0|S_1, \ldots, S_K) &= \prod_{k=1}^{K} P(X_{nm} = 0|S_k) = \prod_{k=1}^{K} \exp(-W_{nk} H_{km}) \\ &= \exp\left(-\sum_k W_{nk} H_{km}\right) \end{aligned} \tag{1.6}$$

This naturally relates the model with *Non-negative Matrix Factorization* (NMF) approaches. The related overall *fail* probability is then given as

$$P(X_{nm} = 1|S_1, \ldots, S_K) = 1 - P(X_{nm} = 0|S_1, \ldots, S_K) = 1 - \exp\left(-\sum_k W_{nk} H_{km}\right) \tag{1.7}$$

This superposition approximation renders chip m classified as *fail* if at least one of the K sources S_k induced a close to zero single event probability $P(X_{nm} = 0|S_k)$. On the contrary, chip m is likely classified as *pass* if, for all $S_k, k = 1, \ldots, K$ the single event probabilities $P(X_{nm} = 0|S_k)$ happen to be large.

Since all W_{nk}, H_{km} are non-negative by assumption, the overall conditional *fail* probability becomes, for small arguments, a linear function of the underlying causes, i.e., $P(X_{nm} = 1|\mathbf{H}_{1*}, \ldots, \mathbf{H}_{K*}) \approx \sum_{k=1}^{K} W_{nk} H_{km}$, and, for large arguments, saturates at $P(X_{nm} = 1|\mathbf{H}_{1*}, \ldots, \mathbf{H}_{K*}) \approx 1$.

1.2.2 Bernoulli Likelihood and Gradient Ascent Optimization

Now we consider the overall *fail* probability (1.7) as a *Bernoulli* variable $\Theta_{nm} = 1 - \exp(-[\mathbf{WH}]_{nm})$. Then the overall *Bernoulli log-likelihood* $\mathscr{L}$ of the binary observation X_{nm} under the Poisson yield model is given by

$$\mathscr{L}(\boldsymbol{\Theta}) = \sum_{n=1}^{N}\sum_{m=1}^{M}\{X_{nm}\ln(\Theta_{nm}) + (1 - X_{nm})\ln(1 - \Theta_{nm})\} \tag{1.8}$$

which has to be maximized in the variables $\mathbf{W}$ and $\mathbf{H}$ with respect to the non-negativity constraints $\mathbf{W}, \mathbf{H} \geq 0$. This constrained optimization problem can be solved conveniently by a basic gradient ascent scheme

$$\begin{gathered} W_{nk} \leftarrow W_{nk} + \eta_W \frac{\partial \mathscr{L}}{\partial W_{nk}} \quad \text{and} \quad H_{km} \leftarrow H_{km} + \eta_H \frac{\partial \mathscr{L}}{\partial H_{km}} \\ \frac{\partial \mathscr{L}}{\partial W_{nk}} = \sum_m H_{km}\left(\frac{X_{nm}}{\Theta_{nm}} - 1\right) \quad \text{and} \quad \frac{\partial \mathscr{L}}{\partial H_{km}} = \sum_n W_{nk}\left(\frac{X_{nm}}{\Theta_{nm}} - 1\right) \end{gathered} \tag{1.9}$$

subject to $W_{nk}, H_{km} \geq 0 \; \forall \; n, m, k$ and where, remember, $\Theta_{nm} = 1 - \exp(-[\mathbf{WH}]_{nm})$.

1.2.2.1 Proper Initialization

The log-likelihood cost function $\mathscr{L}$ given above can induce local optima, thus causing serious global convergence problems. Suppose $X_{nm} = 1$ and the corresponding probability Θ_{nm} is small, this condition will induce a logarithmic divergence of the cost function $\mathscr{L}$. As a consequence, the algorithm will try to compensate this divergent behavior by strongly increasing $[\mathbf{WH}]_{nm}$ as if $[\mathbf{WH}]_{nm} \rightarrow \infty$ then we have $\Theta_{nm} \rightarrow 1$. Such local optima can be suppressed heuristically by observing that for $[\mathbf{WH}]_{nm} \rightarrow \infty$ we have $\Theta_{nm} \approx 1 - \delta$ with δ small. Then we have

$$\Theta_{nm} = \begin{cases} 0 & \text{if } X_{nm} = 0 \\ 1 - \delta & \text{if } X_{nm} = 1 \end{cases} \tag{1.10}$$

Given the binary character of the observations X_{nm}, we can, for all n, m, summarize both conditions such that

$$\begin{aligned} (1 - \delta)X_{nm} &\simeq 1 - \exp([-\mathbf{WH}]_{nm}) \\ \hat{X}_{nm} &:= -\ln(1 - (1 - \delta) \cdot X_{nm}) \simeq [\mathbf{WH}]_{nm} \end{aligned} \tag{1.11}$$

Thus we have the following limiting cases:

$$X_{nm} = 0 \quad \Rightarrow \quad \hat{x}_{nm} = 0 \tag{1.12}$$

$$X_{nm} = 1 \quad \Rightarrow \quad \hat{x}_{nm} = -\ln(\delta) \tag{1.13}$$

Learning proper parameters $[\mathbf{WH}]_{nm}$ can be achieved by optimizing the following Euclidean cost function

$$E(\delta, \mathbf{W}, \mathbf{H}) = \sum_{n=1}^{N} \sum_{m=1}^{M} \left(\hat{X}_{nm} - [\mathbf{WH}]_{nm} \right)^2 \tag{1.14}$$

employing an *Alternating Least Squares* (ALS) algorithm [14] while obeying the constraints $\mathbf{W} \geq 0, \mathbf{H} \geq 0$. The ALS updates are then given by

$$H_{rs} \leftarrow \max\{\varepsilon, [(\mathbf{W}^T\mathbf{W})^{-1}\mathbf{W}^T\hat{\mathbf{X}}]_{rs}\} \tag{1.15}$$

$$W_{lq} \leftarrow \max\{\varepsilon, [\hat{\mathbf{X}}\mathbf{H}^T(\mathbf{H}\mathbf{H}^T)^{-1}]_{lq}\}. \tag{1.16}$$

The choice of the ALS algorithm for this preprocessing to estimate suitable initial values for $[\mathbf{WH}]_{nm}$ is motivated by its speed and simplicity. However, because convergence is not assured, this procedure should be repeated several times using different random initializations of $\mathbf{H}$ and $\mathbf{W}$, and only the solution with the smallest Euclidean distance is retained as a proper initialization to the *binNMF* algorithm (see [1] for a more detailed discussion of this heuristic).

1.2.3 Heuristic Multiplicative Update Rules

Instead of using gradient ascent optimization, an alternative approach would be to consider heuristic multiplicative update rules, where τ represents a characteristic relaxation time of the learning dynamics, typically in the range $[0.5 \leq \tau \leq 2]$.

$$W_{nk} \leftarrow W_{nk} \left[\sum_m \frac{X_{nm}}{\varepsilon + 1 - \exp(-[\mathbf{WH}]_{nm})} \frac{H_{km}}{\sum_m H_{km}} \right]^{1/\tau} \tag{1.17}$$

$$H_{km} \leftarrow H_{km} \left[\sum_n \frac{X_{nm}}{\varepsilon + 1 - \exp(-[\mathbf{WH}]_{nm})} \frac{W_{nk}}{\sum_n W_{nk}} \right]^{1/\tau} \tag{1.18}$$

where $\varepsilon \ll 1$ suppressing singularities. In case of a unit relaxation time, i.e., for $\tau = 1$, it is seen that

- the update rules (1.17) and (1.18) strongly resemble the canonical Lee–Seung [15] updates for NMF.

- these multiplicative update rules can be replaced by the additive update rules (1.9), if we set

$$\eta_W = \frac{W_{nk}}{\sum_m H_{km}} \quad \text{and} \quad \eta_H = \frac{H_{km}}{\sum_n W_{nk}}$$

Such multiplicative updates share with the Lee–Seung updates the benefits of neither having to control a step-size parameter nor a non-negativity constraint, and show good monotonicity properties and a fast convergence. On the other hand, they also share the well-known drawbacks of the Lee–Seung method: convergence toward a saddle point cannot be excluded, and, once a parameter is exactly zero, it remains zero. A general discussion on such multiplicative heuristic formulas can be found in [9].

Concerning the equivalent additive updates, the step-size parameters η_W, η_H are usually small. In this case, the updates resemble a simply gradient ascent step. In case that, for example, H_{km} is large, the gradient is approximately zero, just as in case that $\sum_n W_{nk}$ is very small. It is easy to verify that a local maximum is a fixed point of the above iterations. In the more general case of Eqs. (1.17) and (1.18), one can always choose $\tau \gg 1$ such that the updating factor is close to 1 resulting in small changes between successive values of H_{km}. Note that using the Bernoulli approximation we have

$$\lim_{\tau \to \infty} (1+t)^{\frac{1}{\tau}} \approx \left(1 + \frac{t}{\tau}\right) \quad \text{with } t \ll 1.$$

Hence, the modified update is in effect a gradient ascent. Assuming further that the gradient does not change much among τ successive updates, then τ steps of the modified version are equivalent to one step of the basic multiplicative algorithm (Fig. 1.1).

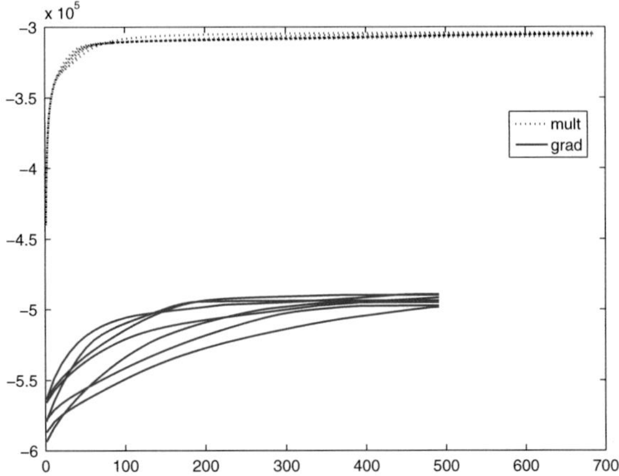

Fig. 1.1 Multiplicative updates versus gradient ascent updates: Both algorithms were initialized by the same random matrices and ran for the same computational time. The multiplicative version shows much better performance

1.2.4 Applications to Wafer Maps

1.2.4.1 Example I

The simulation considers that $K = 6$ basic failure patterns are underlying binary observations from $N = 3043$ wafers, each containing $M = 500$ chips (see Fig. 1.2). The data stems from measurements which aim at identifying latent structures and detecting potential failure causes in an early stage of the processing chain. The estimated $K = 6$ basic failure patterns $\mathbf{H}_{1*}, \ldots, \mathbf{H}_{6*}$ have clearly different characteristics: One pattern of higher *fail* probability on the upper side of the wafer, a bead centered on the wafer, a ring of fails on the edge zone, two different repeated structures, and defects concentrated on the bottom of the wafer were detected. The related weight coefficients $\mathbf{W}_{*1}, \ldots, \mathbf{W}_{*6}$ store the activity of each of the 6 putative

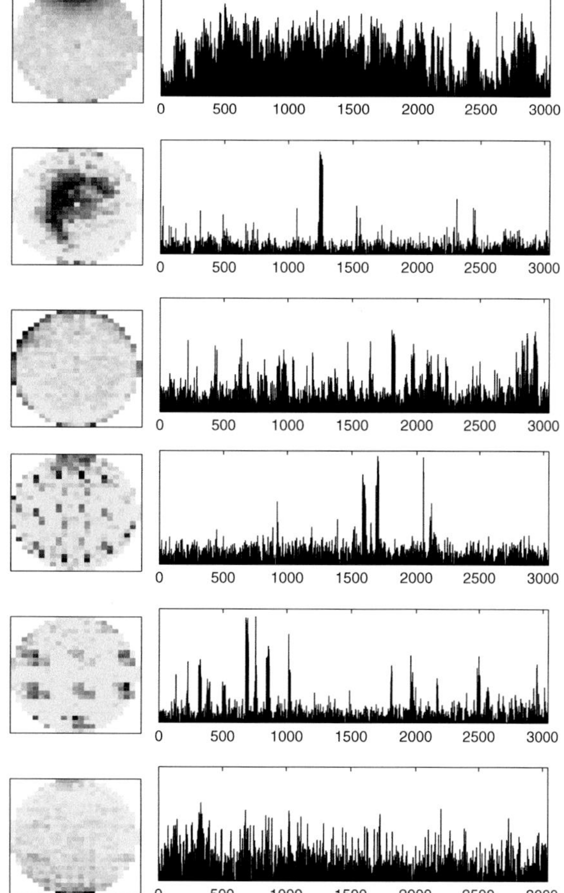

Fig. 1.2 Estimated basic patterns $\mathbf{H}_{1*}, \ldots, \mathbf{H}_{6*}$ and contribution coefficients $\mathbf{W}_{*1}, \ldots, \mathbf{W}_{*6}$, estimated from a real data set comprising $N = 3043$ wafers and $M = 500$ chips per wafer

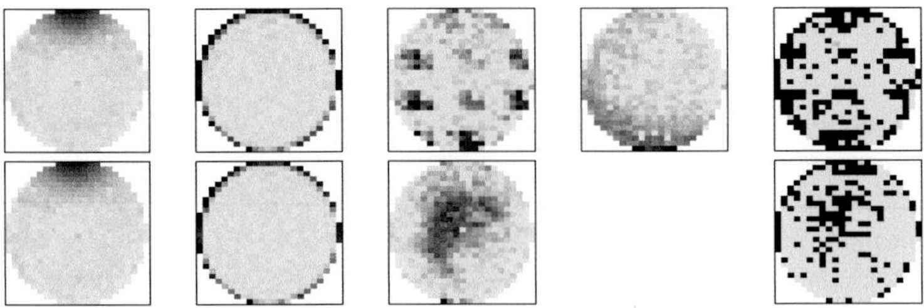

Fig. 1.3 Parts based representation of two different observed wafer maps (*right most images*) as a superposition of underlying patterns of *fail* probabilities $1 - \exp(-W_{nk}H_{k*})$

basic patterns on each wafer separately. This new representation of the data highlights wafers affected by the detected elementary failure patterns and supports the detection of potential error causes (Fig. 1.3).

Figure 1.2 illustrates relevant underlying patterns of *fail* probabilities $1 - \exp(-W_{nk}H_{k*})$ which generate the observed failure patterns. Only patterns are shown which are highlighted by corresponding large weight coefficients W_{nk}.

1.2.4.2 Example II

This second example shows a decomposition of a total of 760 wafers into $K = 6$ source components (see Fig. 1.4). Each wafer contains 3500 chips here and hence the images have a higher resolution. Again, we see different kinds of patterns, some of which seem to have some rotational symmetry. An interesting feature is visible in these images: In wafer manufacturing, a set of typically 25 wafers is called a *lot* and is processed together. However, the process history is not identical for all lots, though it is so for all wafers within one lot. In the example considered here, we can see that the computed source patterns often are active in bundles of about 25 wafers. Either a whole lot is affected by a source or not. Note that the upper leftmost pattern 1 and the middle leftmost pattern 2 seem to be rotated versions of each other and that source 1 becomes active just as source 2 becomes inactive (corresponding to values between 350 and 375 on the abscissa, which represent the indexes of the processed wafers n).

1.2.5 Logistic NMF

The *binNMF* model, employing Bernoulli statistics, can be generalized to a form representing a partially non-negative factorization of a matrix $\boldsymbol{\Theta} = \mathbf{WH}$ of log-odds. The resulting *logistic NMF* model places several constraints onto the factorization process rendering the estimated basis system strictly non-negative or even binary.

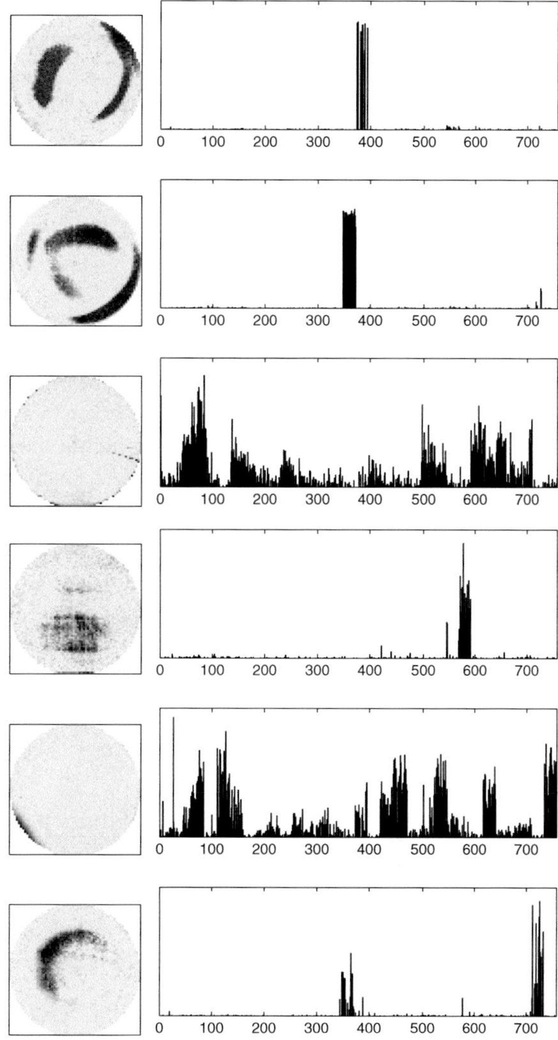

Fig. 1.4 Real-world example II: Estimated basic patterns $\mathbf{H}_{1*}, \ldots, \mathbf{H}_{6*}$ and contribution coefficients $\mathbf{W}_{*1}, \ldots, \mathbf{W}_{*6}$, estimated from a real data set comprising $N = 760$ wafers and $M = 3500$ chips per wafer

Thereby the proposed model places itself in between a *logistic PCA* [16, 17] and a *binary NMF* [1] approach (see Sect. 1.2.1).

As mentioned already, there are various situations [1, 18] where data is binary, e.g., binary wafer maps $\mathbf{X} = (X_{nm} \in \{0, 1\})$ [1]. Suppose that such data can be considered resulting from a transformation of related continuous data $\boldsymbol{\Theta}$ by a unit-step Heaviside function $F(..)$ such that

$$X_{nm} = F(p_{nm}(\theta_{nm}) - 0.5) = \begin{cases} 1 & \text{if } 0.5 \leq p_{nm} \leq 1.0 \\ 0 & \text{if } \ 0 \leq p_{nm} < 0.5 \end{cases} \tag{1.19}$$

Let us further assume that each argument θ_{nm} of the Heaviside function can be decomposed into K latent factors $\mathbf{H}_{k*}$ and their related weights $\mathbf{W}_{n*}$ according to

$$\begin{aligned}\mathbf{X} &= F(\mathbf{P}(\boldsymbol{\Theta}) - 0.5 \cdot \mathbf{1}) \\ \boldsymbol{\Theta} &= \mathbf{WH} \\ \theta_{nm} &= \mathbf{W}_{n*}\mathbf{H}_{*m} \\ \boldsymbol{\Theta}_{n*} &= \sum_k W_{nk}\mathbf{H}_{k*}\end{aligned} \tag{1.20}$$

Thus to each binary observation $\mathbf{X}_{n*}$ corresponds a nonlinearly transformed weighted linear combination of latent or hidden features $\mathbf{H}_{k*}$. This data transformation represents a nonlinear variant of an *Exploratory Matrix Factorization* (EMF) [19] technique where traditionally observed data is interpreted as a linear combination of underlying features characterizing the problem at hand. However, devising a learning strategy and deriving a proper cost function is difficult when employing a Heaviside transformation because of its noncontinuous nature. A continuous and continuously differentiable alternative would be a squashing function like the *logistic function*

$$\sigma(\theta_{nm}) = \frac{1}{1 + \exp[-\theta_{nm}]}. \tag{1.21}$$

Now, let us again model the binary observations X_{nm} by a Bernoulli distribution

$$P(X_{nm}|\theta_{nm}) = (p_{nm}(\theta_{nm}))^{X_{nm}} (1 - p_{nm}(\theta_{nm}))^{(1-X_{nm})} \tag{1.22}$$

We then can relate the occurrence probabilities $p_{nm}(\theta_{nm})$ to the log-odds parameters θ_{nm} via the *logit*-function

$$logit\,(p_{nm}) = \ln\left(\frac{p_{nm}}{1 - p_{nm}}\right) = \theta_{nm} = \mathbf{W}_{n*}\mathbf{H}_{*m}, \tag{1.23}$$

Considering the inverse *logit* function $logit^{-1}(p_{mn})$, i.e., taking the inverse of the logarithm, Eq. (1.23), the Bernoulli parameter p_{nm} can be expressed naturally via the logistic function $\sigma(\theta_{nm})$ as

$$p_{nm} = \frac{1}{1 + \exp[-\theta_{nm}]} = (1 + \exp(-\mathbf{W}_{n*}\mathbf{H}_{*m}))^{-1} = \sigma(\theta_{nm}) \tag{1.24}$$

Note that the Bernoulli variable in case of the *binNMF* approach can be obtained from the above expression in the limit exp(....) $\ll 1$ via the Bernoulli approximation

$$(1 + \exp(-\mathbf{W}_{n*}\mathbf{H}_{*m}))^{-1} \simeq (1 - \exp(-\mathbf{W}_{n*}\mathbf{H}_{*m})) \tag{1.25}$$

To summarize, given the logistic function for the Bernoulli variable p_{nm}, the Bernoulli distribution can be cast in the following form:

$$P(X_{nm}|\theta_{nm}) = \sigma^{X_{nm}}(\theta_{nm})\sigma^{(1-X_{nm})}(-\theta_{nm}) = \frac{1}{1+\exp[-(2X_{nm}-1)\theta_{nm}]} \tag{1.26}$$

where $P(X_{nm}|\theta_{nm})$ measures the probability of X_{nm} exhibiting the value X_{nm}. Under the assumption of independence, the probability distribution over a set of data matrices $\mathbf{X}$ is then given by:

$$P(\mathbf{X}|\theta) = \prod_{n,m} \frac{1}{1+\exp[-(2X_{nm}-1)\theta_{nm}]} \tag{1.27}$$

Remember that the components $\theta_{nm} = \mathbf{W}_{n*}\mathbf{H}_{*m}$, hence the components of the factor matrices $\mathbf{W}$, $\mathbf{H}$, need not be binary, even given binary observations $\mathbf{X}_{nm}$. The logistic function $\sigma(\theta_{nm})$ provides a continuous transformation turning the product of the factor matrices into a matrix of real-valued entries in the range (0, 1). A Heaviside function, if applied at the final stage, would turn these real-valued outcomes of the logistic function into binaries.

1.2.5.1 Gradient-Based Learning

Learning the proposed model can easily be achieved via gradient ascent on a log-likelihood cost function $\mathscr{L}(\mathbf{X}|\mathbf{WH})$ which directly follows from the probability distribution $P(\mathbf{X}|\boldsymbol{\Theta})$ and reads

$$\mathscr{L}(\mathbf{X}|\mathbf{WH}) = -\sum_{n,m} \ln\left(1+\exp\left(-(2X_{nm}-1)[\mathbf{WH}]_{nm}\right)\right) \tag{1.28}$$

Due to the factorization assumption, the entries of the matrix $\boldsymbol{\Theta}$ represent dot products between the row vectors $\mathbf{W}_{n*}$ and the column vectors $\mathbf{H}_{*m}$ of the factor matrices providing the new representation of the relative probabilities. To ensure that the binary data matrix $\mathbf{X}$ be modeled by a logistic function, the only practical constraint needed is having real-valued $\theta_{nm} \in \mathscr{R}$. The resulting values of the logistic function then correspond to probabilities $p_{nm} \geq 0.5$ if $\theta_{nm} \geq 0$, and to probabilities $p_{nm} < 0.5$ if the values for $\theta_{nm} < 0$ are negative. Hence, the binary data X_{mn} is computed according to the probability p_{nm}, such that if $p_{mn} > 0.5$ then $X_{mn} = 1$, otherwise $X_{mn} = 0$. Therefore, in contrary to non-negative matrix factorization, it is not possible to impose a positivity constraint on both factors of the decomposition of $\mathbf{X}$.

Therefore, by regularizing the cost function by integrating two terms which penalize the growth of the norms of the rows $\mathbf{W}_{n*}$ and columns $\mathbf{H}_{*m}$, a new cost function is obtained

$$\mathscr{J}(\mathbf{W},\mathbf{H}) = \mathscr{L}(\mathbf{X}|\mathbf{WH}) - \frac{1}{2}\lambda_W \sum_n \|\mathbf{W}_{n*}\|^2 - \frac{1}{2}\lambda_H \sum_m \|\mathbf{H}_{*m}\|^2 \tag{1.29}$$

where λ_W, λ_H represent scalar Lagrangian parameters. The corresponding derivative of the regularized cost function with respect to each element of matrix $\mathbf{W}$ then reads

$$\frac{\partial \mathscr{J}(\mathbf{W},\mathbf{H})}{\partial W_{n'k'}} = \sum_m \frac{(2X_{n'm}-1)}{1+\exp\left((2X_{n'm}-1)[\mathbf{WH}]_{n'm}\right)} H_{k'm} - \lambda W_{n'k'} \tag{1.30}$$

Note that these penalties render the update rule dependent on the previous value of the related element of the factor matrix. The global update rule can now be obtained and, in matrix notation, is given by

$$\mathbf{W} \leftarrow \mathbf{W} + \eta_W \nabla_{\mathbf{W}} \mathscr{J}(\mathbf{W},\mathbf{H}) = \mathbf{W} + \eta_W \Delta\mathbf{W} \tag{1.31}$$

$$\Delta\mathbf{W} = \frac{2\mathbf{X}-\mathbf{1}}{\mathbf{1}+\exp\left((2\mathbf{X}-\mathbf{1})\bigodot(\mathbf{WH})\right)}\mathbf{H}^T - \lambda_W \mathbf{W} \tag{1.32}$$

$$\mathbf{H} \leftarrow \mathbf{H} + \eta_H \nabla_{\mathbf{H}} \mathscr{J}(\mathbf{W},\mathbf{H}) = \mathbf{H} + \eta_H \Delta\mathbf{H} \tag{1.33}$$

$$\Delta\mathbf{H} = \mathbf{W}^T \frac{2\mathbf{X}-\mathbf{1}}{\mathbf{1}+\exp\left((2\mathbf{X}-\mathbf{1})\bigodot(\mathbf{WH})\right)} - \lambda_H \mathbf{H} \tag{1.34}$$

Note that here $\bigodot$ denotes the Hadamard product, i.e., a pointwise product between the elements of the two matrices, and $\mathbf{1}$ is the matrix with all elements being one. Also the division and the exponential function have to be understood as pointwise operations. For simplicity, $\lambda_W = \lambda_H = \lambda$ is taken.

Non-negative matrix factorization techniques [20] often apply a projection step to confine the matrix elements to the positive orthant. Using the logistic function to model the occurrence probabilities of the binary data, such a projection step can be applied only to one of the factor matrices, as explained above, thus representing a nonlinear semi-NMF approach.

1.2.5.2 Post-processing Step

An alternative way to transform real-valued model outputs θ_{nm} into binary observations $\tilde{x}_{nm} \in \{0, 1\}$ is using the sign function sgn $(\theta_{mn}) =$ sgn $(\mathbf{W}_{n*}\mathbf{H}_{*m})$. This also allows to put stricter constraints to the matrix factors estimated by the algorithm. One possibility is to restrict the basis vector matrix $\mathbf{H}^T$ to contain binary entries only. Such binarization of $\mathbf{H}^T$ could be achieved in a post-processing step, for example. Considering the basic assumption of the model that the basis vector matrix $\mathbf{H}^T$ should have only entries $h^T_{nd} \geq 0$, an iterative post-processing manipulation could substitute the L largest values of its columns by 1, while the remaining values are assigned to 0. The dimension L should be chosen to assure a minimal number of mismatches between the original data $\mathbf{X}$ and the reconstructed data $\tilde{\mathbf{X}}^{(\mathbf{L})}$ according to

$$L^* = \arg\min_L \|\mathbf{X} - \tilde{\mathbf{X}}^{(\mathbf{L})}\|_F^2$$

where $\|...\|_F$ denotes the Frobenius norm. An optimal $L*$ and corresponding binary matrix $(\tilde{\mathbf{H}}^{\mathbf{T}})^{(L)}$ can be computed iteratively as follows:

1. Initialize the maximum number of possible mismatches to $e(0) = MN$ and set $L = 1$.
2. Compute a binary matrix $(\tilde{\mathbf{H}}^{\mathbf{T}})^{(L)}$ substituting the L largest entries of each column by ones and the remaining entries by zeros.
3. Reconstruct the binary data $\tilde{\mathbf{X}}^{(\mathbf{L})}$ according to

$$\tilde{x}_{nm}^{(L)} = F(p_{nm} - 0.5) = F\left(\sigma(\mathbf{W}_{n*}\mathbf{H}_{*m}^{L}) - 0.5\right)$$

4. Compute the mismatch error

$$e(L) = \|\mathbf{X} - \tilde{\mathbf{X}}^{(\mathbf{L})}\|_F^2$$

5. If number of mismatches increases, then stop and

   - accept the basis vector matrix of the previous iteration, i.e., $\tilde{\mathbf{H}}^{\mathbf{T}}$ with the first $L - 1$ entries of each column replaced by 1, as the solution.
   - Otherwise $L \leftarrow L + 1$ and go to step 2.

1.2.5.3 An Illustrative Toy Example

The application of the proposed learning rules necessitates model parameters to be preassigned: the internal dimension of the factorization, i.e., the number (K) of columns (rows) of the factors $\mathbf{W}$ $(\mathbf{H})$, the momentum rate (λ), and naturally the learning rate (η) of the updating rules: $\mathbf{W}^{(t+1)} = \mathbf{W}^{(t)} + \eta\nabla\mathbf{W}$ and $\mathbf{H}^{(t+1)} = \mathbf{H}^{(t)} + \eta\nabla\mathbf{H}$. These rules will be applied following an alternating scheme in each tth iteration and in the following simulations $\lambda = 0.1$ and the total number of iterations was set to 1000.

The original binary data matrix $\mathbf{X}$ was created according to Eq. (1.19) where the matrix $\mathbf{W}^T$ has Gaussian random entries with zero mean and unitary standard deviation and the columns of the matrix $\mathbf{H}^T$ are formed by sparse binary images [18]. Therefore, 16 distinct vertical and horizontal bar images (8×8) [18] are concatenated to form the columns of matrix $\mathbf{H}^T$. The simulations illustrate the performance of the algorithm before and after the binarization step. There 8 different values of learning rates were used and for each the dimension (K) of the factors was assigned to: 12, 16 (the actual value), and 20.

The alternating gradient ascent update rule finds solutions for both factor matrices $\mathbf{W}$ and $\mathbf{H}$ that lead to a very small (or even zero) reconstruction error if the internal dimension (K) of the matrix factorization is overestimated or correctly estimated. Tables 1.1 and 1.2 show the average and standard deviation of the reconstruction

Table 1.1 Reconstruction error (mean and standard deviation) after 1000 iterations for different learning rates (η)

Experiment	η	$K = 12$	$K = 16$	$K = 20$
1	0.0005	0.35 ± 0.006	0.34 ± 0.006	0.33 ± 0.007
2	0.001	0.08 ± 0.005	0.03 ± 0.006	0.01 ± 0.002
3	0.002	0.05 ± 0.003	0	0
4	0.005	0.04 ± 0.003	0	0
5	0.01	0.04 ± 0.002	0	0
6	0.02	0.04 ± 0.003	0	0
7	0.05	0.07 ± 0.004	0	0
8	0.07	0.1 ± 0.004	0	0

Table 1.2 Reconstruction error (mean and standard deviation) after the post-processing step

Experiment	$K = 12$	$K = 16$	$K = 20$
1	0.33 ± 0.013	0.32 ± 0.02	0.31 ± 0.012
2	0.18 ± 0.006	0.14 ± 0.006	0.11 ± 0.006
3	0.19 ± 0.007	0.13 ± 0.006	0.11 ± 0.005
4	0.21 ± 0.010	0.14 ± 0.006	0.11 ± 0.006
5	0.22 ± 0.010	0.13 ± 0.006	0.11 ± 0.005
6	0.21 ± 0.011	0.13 ± 0.007	0.11 ± 0.008
7	0.16 ± 0.008	0.08 ± 0.003	0.07 ± 0.019
8	0.15 ± 0.006	0.04 ± 0.02	0.04 ± 0.014

error before and after binarization, respectively. Applying the post-processing step to binarize the entries of matrix $\tilde{\mathbf{H}}^T$, the reconstruction error increases. However, the original bar patterns can be identified easily in the solutions achieved (see Fig. 1.5). With $K = 16$, the original patterns are recovered exactly. Figure 1.6 illustrates the dependence of the number of nonzero entries in each column of the matrix $\tilde{\mathbf{H}}^T$ of basis vectors of the new representation.

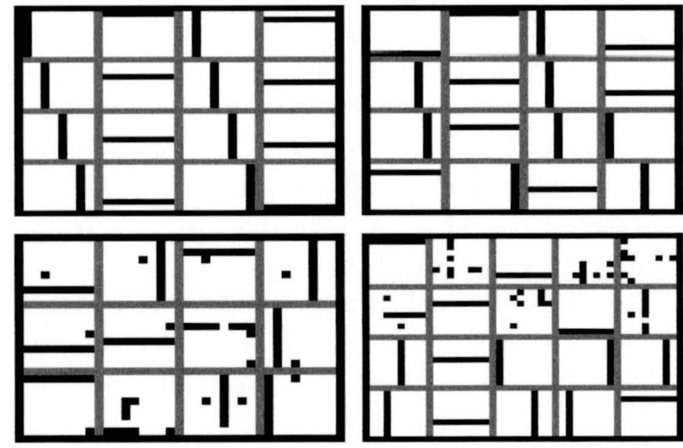

Fig. 1.5 The binary image data set. The *top left* image represents the original data set. Each other image illustrates the set of columns of $\tilde{\mathbf{H}}^T$: *top right*-$K = 16$; *bottom left*-$K = 12$; *bottom right*-$K = 20$. For visualization purposes, *black* represents one and *white* zero

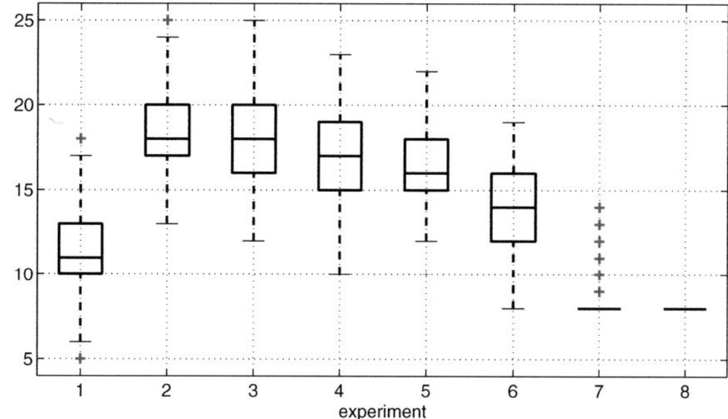

Fig. 1.6 Post-processing step ($K = 20$): the number of nonzero in the matrix $\tilde{\mathbf{H}}^T$ for each of the solutions of the gradient rules. Each *box* represents eight experiments

1.2.6 Related Works for Binary Data Sets

Two other models are discussed in the literature which utilize the Bernoulli likelihood to model binary data via real-valued source and weight variables. They are the *logistic PCA* model [16, 21], and the *Aspect Bernoulli* model [22]. A brief discussion of both puts them into perspective to the model proposed in this study.

1.2.6.1 Logistic PCA

Logistic PCA is an extension of the canonical principal component analysis (PCA) to deal with a dimension reduction of binary data sets. We here summarize under the term *logistic PCA* the works of [16, 21, 23] which are closely related to each other. Tipping [21] aimed at visualizing binary data sets and used a variational technique to determine optimal model parameters. Collins et al. [23] extended PCA to the exponential family for providing dimensionality reduction schemes to non-Gaussian distributions like the Bernoulli distribution. The authors describe a general optimization scheme in terms of Bregman divergences. Applying their framework to binary data sets thus leads to the model likelihood given below Eq. (1.35). Schein et al. [16] used a method similar to [21] which they called *logistic PCA* and demonstrated its potential for data reconstruction on a variety of real world problems.

Logistic PCA can be formulated as an optimization problem of the following likelihood function $\mathscr{L}(\mathbf{W}, \mathbf{H})$

$$\mathscr{L}(\mathbf{W}, \mathbf{H}|X_{nm}) \equiv P(X_{nm}|\mathbf{W}, \mathbf{H}) = \sigma([\mathbf{WH}]_{nm})^{X_{nm}}(1 - \sigma([\mathbf{WH}]_{nm}))^{1-X_{nm}} \tag{1.35}$$

where $\sigma(z) = (1 + \exp(-z))^{-1}$ is the logistic function. *Logistic PCA* and *binNMF* both optimize a Bernoulli likelihood which is parameterized by the set of Bernoulli parameters $\Theta_{nm} \in [0, 1]$. Both methods utilize a nonlinear transfer function which maps the matrix product $\mathbf{WH}$ onto the range of the Θ_{nm}. The main difference is given by the non-negativity constraints imposed in our *binNMF* model. By imposing no non-negativity constraints on the factor matrices **W** and **H** in the *logistic PCA* model, negative solutions can be obtained. The latter have no meaning in the context of a strictly additive superposition of non-negative source variables which forms a basic characteristic of our model.

As a nonlinear version of PCA, *logistic PCA* has inherited most of its advantages and drawbacks. While being very powerful in dimension reduction applications, the new representation has often no intuitive interpretation within the underlying model. This issue clearly has analogies to the case of continuous variables, where positive [6] or non-negative [7, 8] matrix factorization, respectively, finds a representation whose components have an immediate and intuitive interpretation within the model in contrast to a representation resulting from a linear PCA transformation.

1.2.6.2 Aspect Bernoulli

The *Aspect Bernoulli* model, as introduced by Kaban et al. [22], and Bingham et al. [24], does not need any nonlinearity. It decomposes the mean of the Bernoulli distribution $P(X_{nm} = 0)_{AB} = [\mathbf{WH}]_{nm}$ instead of $P(X_{nm} = 0) = 1 - \exp(-[\mathbf{WH}]_{nm})$, thus yielding the likelihood function

$$\mathscr{L}(\mathbf{W}, \mathbf{H}|X_{nm}) \equiv P(X_{nm}|\mathbf{W}, \mathbf{H}) = [\mathbf{WH}]_{nm}^{X_{nm}} (1 - [\mathbf{WH}]_{nm})^{1-X_{nm}}. \qquad (1.36)$$

The *Aspect Bernoulli* model constrains the entries of both parameter matrices **W** and **H** to values $W_{nk}, H_{km} \in [0, 1]$. Here H_{km} stores the Bernoulli probability of the mth attribute being *on*, conditioned on the latent aspect k, while W_{nk} is the probability of choosing a latent aspect k in observation n. The entries of row $\mathbf{W}_{n*}$ are restricted to sum to 1, thus keeping $[\mathbf{WH}]_{nm} \leq 1$ like is done in a standard mixture model. This key feature, however, renders the *Aspect Bernoulli* model not suitable for the analysis of wafer maps as discussed in Sect. 1.2.1. Cases where several sources are highly active simultaneously cannot be displayed properly. Suppose there are two aspects (called elementary causes in our model) highly active in one observation and each aspect has a probability larger than 0.5. This would yield $W_{n1}H_{1m} > 0.5$ and $W_{n2}H_{2m} > 0.5 \Rightarrow WH_{nm} > 1$, contradicting $[\mathbf{WH}]_{nm} \in [0, 1]$. Hence, such cases cannot be handled by the *Aspect Bernoulli* model.

Bingham et al. [24] present a rather informative overview of related models and even mention that an intermediate model between *logistic PCA* and *Aspect Bernoulli* could be constructed. The *binNMF* model is an example of such an intermediate model which is to be positioned right between the *logistic PCA* model and the *Aspect Bernoulli* model.

1.3 Probabilistic Approaches to NMF

Though dealing with matrix factorization issues, up till now we put aside questions about how we can assure having unique solutions to the matrix factorization problem. Also we tacitly assumed firm knowledge about the intrinsic dimension K of the problem. In this section we now address uniqueness and model order selection issues within a probabilistic treatment. First, we consider certain shortcomings of considering NMF as an optimization approach given non-negativity constraints only. We argue that obtaining unique solutions and robustly estimating the intrinsic dimension of a data set affords additional constraints or a more principled approach provided by a Bayesian probabilistic treatment.

Bayesian probability theory offers a convenient way to incorporate prior knowledge about parameter distributions [25]. Bayesian techniques provide means for determining optimal parameter settings through maximizing data likelihood. But they also provide a framework for model comparison [26]. In NMF settings, different models can be identified by their different numbers of underlying modes K. Assuming a model $\mathscr{M}_K$, the posterior distribution of parameters times the evidence which the data provides for model $\mathscr{M}_K$ equals the likelihood multiplied by the prior of the parameters according to the famous Bayes' rule. Thus, we approach a probabilistic treatment of NMF in a two step process:

- At a first level of Bayesian inference, the most probable set of parameters $\mathbf{W}, \mathbf{H}$ is estimated by maximizing the likelihood $P(\mathbf{W}, \mathbf{H}|\mathbf{X}, K)$ given model $\mathscr{M}_K$ with a *fixed* number K of relevant features. *Maximum Likelihood Estimation* (MLE) can be extended, however, if prior knowledge about model parameters is included resulting in an estimation of their posterior distributions. Contrasting the latter strategy, we propose a *Bayesian Optimality Criterion* (BOC) for NMF solutions [12], which can be derived even in the absence of prior knowledge about the latent factors. It is based on a Euclidean distance measure and leads to a minimum volume constraint, thus connecting our geometrical view of NMF with a Bayesian point of view.
- At a second level of inference, we then compare alternative models with each other by evaluating the evidence for model $\mathscr{M}_K$ for different model orders K. We present a *Variational Bayes NMF* algorithm *VBNMF* [4] for optimizing model complexity and tackling the problem of model order selection. This *VBNMF* approach will interpret different models by different factorization ranks K. We will see how Bayesian inference can help to determine the optimal number of modes in a given data set under specific prior assumptions. We show that *VBNMF* is a straightforward generalization of the canonical Lee–Seung method [8, 13] for the Euclidean NMF problem. We further demonstrate its ability to automatically detect the number of intrinsic non-negative components underlying non-negative observations. We also explore the potential of the VBNMF algorithm in applications to binary data sets, thus generalizing the approaches discussed above. Finally, we discuss our approach in light of other Bayesian NMF methodologies reported in literature.

1.3.1 NMF as a Constrained Optimization Problem

Remember that the matrix factorization problem posed by NMF can be converted into an optimization problem by defining a proper cost function. The $(N \times M)$-dimensional non-negative data matrix $\mathbf{X}$ of observations is approximated by the product of an $(N \times K)$-dimensional weight matrix $\mathbf{W}$ and a $(K \times M)$-dimensional matrix $\mathbf{H}$ of hidden features according to the data model discussed above. Obviously, the matrix of residual $\mathbf{E}$ tends to have small entries $E_{nm} = X_{nm} - [\mathbf{WH}]_{nm}$ if the product $\mathbf{WH}$ approximates the data $\mathbf{X}$ well. Note that the statistics of the residuals E_{nm} specify the type of cost function which should be optimized. A commonly used cost function, referring to normally distributed residuals, is the quadratic error function D_E

$$D_E(\mathbf{X}, \mathbf{WH}) = \frac{1}{2} \sum_n \sum_m (X_{nm} - [\mathbf{WH}]_{nm})^2 \tag{1.37}$$

subject to the constraint that either factor matrix has only non-negative entries, i.e., $\mathbf{W} \geq 0$, $\mathbf{H} \geq 0$. The parameter matrices $\mathbf{W}$ and $\mathbf{H}$ are initialized properly, and iteratively updated in order to decrease the cost function while keeping the non-negativity constraints valid.

The cost function (1.37) is just the most popular choice for NMF. It is based on the Frobenius norm of the reconstruction error of the data matrix. This Euclidean distance measure is proper for a Gaussian error model. Other, less popular, cost functions refer to different error statistics and are based mostly on information theoretic concepts like the generalized Kullback–Leibler divergence, the Itakura–Saito divergence or similar divergences [27–29] (see [9, 30] for a review).

Besides technical issues, such as a fast implementation and concepts to handle large data sets in times of *Big Data* [5], there are two issues which call for additional constraints when formulating NMF as an optimization problem (similar to Eq. (1.37)) and need further discussion:

- First, even in the noise-free case, i.e., when $\mathbf{X} = \mathbf{WH}$ holds exactly, the constraints $\mathbf{W}, \mathbf{H} \geq 0$ are not sufficient to determine a unique solution. Since $\mathbf{WH} = \mathbf{WS}^{-1}\mathbf{SH}$ for any invertible matrix $\mathbf{S}$, any pair $(\mathbf{WS}^{-1}, \mathbf{SH})$ still provides a valid decomposition of the same data $\mathbf{X}$ as long as $\mathbf{S}^{-1}$ exists and $\mathbf{WS}^{-1}, \mathbf{SH} \geq 0$. So, applying an NMF algorithm which optimizes Eq. (1.37) twice with two different initializations of $\mathbf{W}$ and $\mathbf{H}$ for a given K will in general yield two different solutions. Furthermore, due to the structure of most cost functions, gradient-based algorithms can get stuck in local optima. Thus, in practice, different initializations yield different solutions mostly. A geometric interpretation of NMF helps to grasp the essence of the problem and will be considered next.
- Second, the number K of latent features, which are represented by the rows of $\mathbf{H}$ and their contributions to the observations which are collected in the entries of the related row of $\mathbf{W}$, is required to be known in advance. In real-world applications such knowledge is usually missing.

1.3.1.1 A Geometric View on NMF

Here we approach the NMF problem from a geometrical point of view and discuss uniqueness issues in case of negligible noise contributions. In the following, we concentrate on a noise-free situation, i.e., $E_{nm} = 0,\ \forall\, n, m$. We further denote with $\mathbf{X}_{n*}$ the nth row of matrix $\mathbf{X}$. In this notation, we assume the M-dimensional data $\mathbf{X}_{n*},\ n = 1, \ldots, N$ to be generated by non-negative linear combinations of $K \leq M$ linearly independent feature vectors $\mathbf{H}_{k*} \geq 0$ such that

$$\mathbf{X}_{n*} = \mathbf{W}_{n*}\mathbf{H}. \tag{1.38}$$

We further assume, without loss of generality, the row vectors $\mathbf{H}_{k*}$ to be normalized such that $\sum_m (H_{km})^2 = 1$ for $k = 1, \ldots, K$. Since we have

$$X_{nm} = \sum_k W_{nk} H_{km} = \sum_k (W_{nk}\alpha_k)(\alpha_k^{-1} H_{km}), \tag{1.39}$$

such a normalization can be achieved always by rescaling the columns of the weight matrix $\mathbf{W}_{*k}$ and the feature vectors $\mathbf{H}_{k*}$ by setting $\alpha_k = \sqrt{\sum_m (H_{km})^2}$.

Geometrically, non-negativity manifests itself as follows: see Fig. 1.7 for a three-dimensional example where $M = 3,\ K = 2$, see also [31–33]

- $\mathbf{X} \geq 0$:
  All data vectors $\{\mathbf{X}_{i*}\}_{i=1}^N$ constitute a cloud in the non-negative orthant $(\mathbb{R}_0^+)^M$ of $\mathbb{R}^M$.

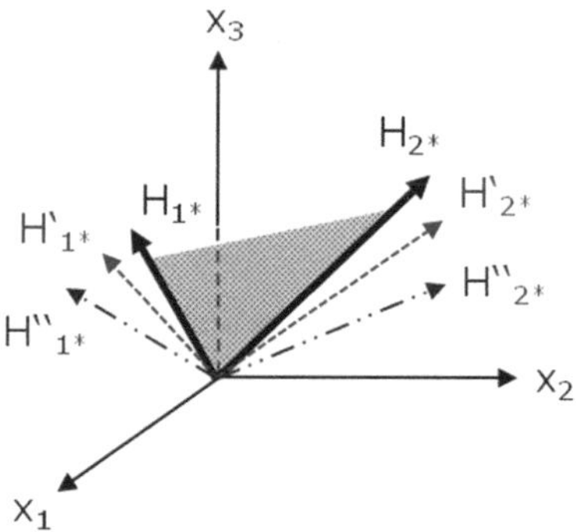

Fig. 1.7 Three-dimensional geometrical interpretation of the NMF problem $\mathbf{X} = \mathbf{W}\mathbf{H}$. The $M = 3$-dimensional data $\mathbf{X}$ lie in the positive span of $K = 2$ basis vectors in the rows of $\mathbf{H}$. The coordinate axes are given by three unit vectors $\mathbf{X_1}$, $\mathbf{X_2}$, $\mathbf{X_3}$; The data lies inside a two-dimensional region between the basis vectors $\mathbf{H}_{1*}$, $\mathbf{H}_{2*}$, but also between the alternative basis vectors $\mathbf{H}'_{1*}$, $\mathbf{H}'_{2*}$ and $\mathbf{H}''_{1*}$, $\mathbf{H}''_{2*}$, respectively

- $\mathbf{H} \geq 0$:
  Each basis vector $\mathbf{H}_{k*}$ points into $(\mathbb{R}_0^+)^M$; normalized vectors $\mathbf{H}_{k*}$ span a K-dimensional subset of the non-negative orthant $(\mathbb{R}_0^+)^M$ of $\mathbb{R}^M$. For example, if $K = 3$, the boundaries of this subset represent three edges of a tetrahedron, intersecting at the origin.
- $\mathbf{W} \geq 0$:
  Positive coefficients W_{nk} for all k imply that a data vector, represented by $\mathbf{X}_{n*} = \mathbf{W}_{n*}\mathbf{H}$, lies inside this K-dimensional subset, whereas $W_{nk} = 0$ for at least one k indicates that point X_{n*} lies on a K-1 dimensional peripheral surface. Here, we consider the surfaces as part of the interior of the subset.

Consequently, any set of K non-negative vectors $\{\mathbf{H}_{k*}\}_{k=1}^{K}$ forming a convex hull which totally encloses the data $\mathbf{X}$, provides a valid solution to the NMF problem $\mathbf{X} = \mathbf{WH}$.

Given now a solution, $\mathbf{W}$, $\mathbf{H}$, and a linear mapping $\mathbf{S}$ which transforms the matrix $\mathbf{H} \mapsto \mathbf{SH}$, then the resulting vectors $(\mathbf{SH})_{k*}$ provide another solution, if all $(\mathbf{SH})_{k*} \geq 0$ **and** all data lie inside the new parallelepiped ($\Leftrightarrow \mathbf{WS}^{-1} \geq 0$), since

$$\mathbf{X} = \mathbf{WH} = \mathbf{WS}^{-1}\mathbf{SH} \tag{1.40}$$

As Fig. 1.7 illustrates, non-negativity constraints are rarely sufficient to define intrinsic features uniquely in the general case. In [3, 11], we proposed an additional minimal determinant constraint and developed the algorithm *detNMF*. Within the context of NMF, such minimum volume constraints were developed independently in the remote sensing community [34], where minimum volume concepts were used to recover source signals from mixed signals since 1994 [35].

In summary, such undesired variability of potential NMF solutions is pointing toward a need for additional constraints. Technically, such constraints can be realized via additional penalty terms in the NMF cost function to be optimized like

$$D_{constr}(\mathbf{X}, \mathbf{WH}) = \frac{1}{2}\sum_n \sum_m (X_{nm} - [WH]_{nm})^2 + \lambda_W f(\mathbf{W}) + \lambda_H g(\mathbf{H}). \tag{1.41}$$

Here, the scalars λ_W, λ_H determine the balance between reconstruction accuracy and the desired properties of the factor matrices, and the functions $f()$ and $g()$ constrain the solutions to these desired properties of the factor matrices. Such approaches have been discussed in literature frequently, mainly based on sparseness considerations. Figure 1.8, however, illustrates a situation where sparseness constraints fail to achieve appropriate solutions contrary to detNMF (see [3] for further discussion). A more principled approach toward achieving unique solutions, i.e., handling uncertainty, and incorporating prior knowledge into NMF problems is offered by applying Bayesian techniques which we will discuss next.

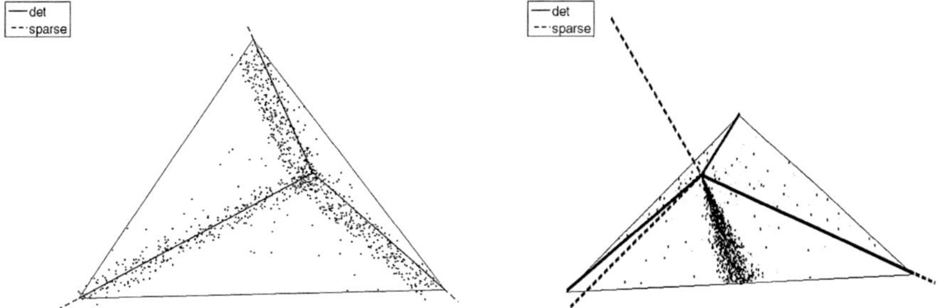

Fig. 1.8 Feature vectors recovered by a volume constraint algorithm *detNMF* (*solid lines*) and by a NMF algorithm with sparsity constraint *nnsc*. *Left* sparse data example, e.g., most of the data are located next to the feature vectors. *Right* highly mixed data example, e.g., most of the data are located between two feature vectors

Introduction to Bayesian NMF

In NMF settings, we can associate different models with different numbers of underlying sources K. Assuming a model $\mathcal{M}_K$ with complexity K, the posterior distribution of parameters times the evidence which the data provide for model $\mathcal{M}_K$ equals the likelihood multiplied by the prior according to Bayes' rule [26, 36]

$$P(\mathbf{W}, \mathbf{H}|\mathbf{X}, K)P(\mathbf{X}|K) = P(\mathbf{X}|\mathbf{W}, \mathbf{H})P(\mathbf{W}, \mathbf{H}|K) \quad (1.42)$$

In a first level of Bayesian inference, the most probable set of parameters $\mathbf{W}, \mathbf{H}$, given a fixed number K of relevant features, can be estimated by maximizing the posterior

$$P(\mathbf{W}, \mathbf{H}|\mathbf{X}, K) = \frac{P(\mathbf{X}|\mathbf{W}, \mathbf{H})P(\mathbf{W}, \mathbf{H}|K)}{P(\mathbf{X}|K)} \quad (1.43)$$

w.r.t. the parameters $\mathbf{W}$ and $\mathbf{H}$. Depending on the specific choice of likelihood function and prior distributions, various well-known NMF problems can be seen as maximum likelihood or maximum a posteriori estimations.

For example, assuming $P(\mathbf{W}, \mathbf{H}|K) = P(\mathbf{W}|K)P(\mathbf{H}|K)$ and taking the logarithm on both sides yields

$$\ln P(\mathbf{W}, \mathbf{H}|\mathbf{X}, K) = \ln P(\mathbf{X}|\mathbf{W}, \mathbf{H}) + \ln P(\mathbf{W}|K) + \ln P(\mathbf{H}|K) - \ln P(\mathbf{X}|K) \quad (1.44)$$

Now using a Gaussian likelihood function with spread parameter σ_r

$$P(\mathbf{X}|\mathbf{W}, \mathbf{H}) = \prod_n \prod_m \frac{1}{\sqrt{2\pi}\sigma_r} \exp\left(-\frac{1}{2}\left(\frac{X_{nm} - [\mathbf{W}\mathbf{H}]_{nm}}{\sigma_r}\right)^2\right) \quad (1.45)$$

yields

$$\ln P(\mathbf{X}|\mathbf{W}, \mathbf{H}) = -NM \ln(\sqrt{2\pi}\sigma_r) - \frac{1}{2\sigma_r^2} \sum_n \sum_m (X_{nm} - [\mathbf{W}\mathbf{H}]_{nm})^2 \tag{1.46}$$

which is, up to constants w.r.t. **W** and **H**, the negative of the squared Euclidean distance measure for NMF (1.37). Now adding prior knowledge by choosing, for example, exponential priors of the form

$$P(\mathbf{W}|K) = \prod_n \prod_k \lambda \exp(-\lambda W_{nk}) \tag{1.47}$$

with $\lambda > 0$ leads to

$$\ln P(\mathbf{W}|K) = -\lambda \sum_{nk} W_{nk} \tag{1.48}$$

which constitutes the additional penalty terms used in [37] to enforce sparse coding of the weight matrix **W**. Thus, non-negative sparse coding can be interpreted as a *maximum a posteriori* (MAP) estimation by maximizing Eq. (1.44) w.r.t. **W** and **H**, assuming independent exponential prior distributions of the weights W_{nk} and flat non-negative priors on the features H_{km} together with a Gaussian likelihood function as in Eq. (1.45). Note that is possible to chose $P(\mathbf{H}|K)$ sufficiently flat but nonzero only in the non-negative range, so that the terms $\ln P(\mathbf{H}|K)$ become irrelevant for the maximization of Eq. (1.44). Since the last term $\ln P(\mathbf{X}|K)$ does not depend on **W** and **H** it can be dropped as well.

1.3.2 Estimating Posterior Distributions of Latent Factors Without Prior Knowledge

Thus at a very basic level, Bayesian inference amounts to maximizing the likelihood $\mathcal{L}(\boldsymbol{\Theta}) \equiv P(\mathbf{X}|\boldsymbol{\Theta})$ of a set of parameters $\boldsymbol{\Theta}$ given a set of observations **X** and an underlying data model $\mathcal{M}_K$ with K intrinsic modes. Above, NMF was introduced in terms of optimizing a suitable cost function subject to non-negativity constraints. However, many popular NMF cost functions can be related to statistical models via *Maximum Likelihood* (ML) estimations. In other words, the statistics reflected by the distribution of reconstruction residuals determines the type of cost function appropriate for the factorization problem at hand. For example, the squared Euclidean distance measure used above (Eq. 1.37) is based on Gaussian error statistics, while

KL- or IS divergences as cost functions relate to alternative error statistics given by Poisson or Gamma distributed noise kernels, respectively (see, e.g., [20, 32, 38–40]). At a slightly more advanced level, Bayesian techniques can be used to explicitly incorporate prior knowledge about the factor matrices by which the matrix of observations is to be approximated. Such priors include independent Gamma priors [41], Gaussian process priors [38], or Gamma chain priors [42] for audio signal modeling and allow to estimate corresponding posterior distributions $P(\boldsymbol{\Theta}|\mathbf{X})$. A common feature of all such approaches is that they seek to model the characteristics of *desired* solutions by *specifically tuned* shapes or properties of the prior distributions $P(\boldsymbol{\Theta})$.

Contrasting such approaches, we will show how Bayesian techniques can also cope with cases where additional information on the properties of the factor matrices is missing and derive a Bayesian optimality criterion for NMF solutions in Sect. 1.3.2.1. We will see that this approach is closely related to minimum volume constraints imposed on the optimization of proper cost functions [11] while tackling the uniqueness problem of NMF.

1.3.2.1 Bayesian Optimality Condition for NMF with Gaussian Likelihood

We have seen in Sect. 1.3.1 that the non-negativity constraints on $\mathbf{W}$ and $\mathbf{H}$ are insufficient to identify the global optimum of a given NMF problem. In the spirit of the first level of Bayesian inference, we are now going to formulate a *Bayesian Optimality Condition* (BOC) for estimating a set of features, represented by the rows $\mathbf{H}_{k*}$, $k = 1, \ldots, K$ of matrix $\mathbf{H}$, which optimally explain the set of observations $\mathbf{X}$ under the prerequisite of Gaussian error statistics and a known model complexity. Given non-negative data $\mathbf{X}$ and the number K of underlying features but no additional prior information, we are interested in the most probable matrix $\mathbf{H} \geq 0$ representing the observations $\mathbf{X} \geq 0$ via weights $\mathbf{W} \geq 0$. This estimation requires the computation of the posterior distribution $P(\mathbf{H}|\mathbf{X}, K) \propto P(\mathbf{X}|\mathbf{H}, K)P(\mathbf{H}|K)$, which can be deduced from the associated data likelihood $P(\mathbf{X}|\mathbf{H}, K)$ and related priors $P(\mathbf{H}|K)$ according to Bayes' rule. Since K is assumed to be known a priori here, we omit it in the formulas and write, e.g., $P(\mathbf{X})$ instead of $P(\mathbf{X}|K)$ for the sake of simplicity. Expressing our ignorance about the distribution of $\mathbf{H}$, the posterior can be obtained from the related data likelihood. We show that the corresponding high-dimensional integral can be solved approximately by employing Laplace's approximation.

According to Bayes' rule, the posterior distribution of the feature vectors, given the data, can be expressed as

$$P(\mathbf{H}|\mathbf{X}) = \frac{P(\mathbf{X}|\mathbf{H})P(\mathbf{H})}{P(\mathbf{X})} \tag{1.49}$$

Here, $P(\mathbf{X})$ can be considered a normalizing constant, $P(\mathbf{H})$ denotes the prior on the feature vectors and the term

$$P(\mathbf{X}|\mathbf{H}) = \int P(\mathbf{X}|\mathbf{W}, \mathbf{H}) P(\mathbf{W}) d\mathbf{W} \tag{1.50}$$

represents the likelihood integrated over all possible weights W_{nk}, hence involves NK dimensions.

Taking logarithm on both sides of Eq. (1.49) yields

$$\ln P(\mathbf{H}|\mathbf{X}) = \ln \int P(\mathbf{X}|\mathbf{W}, \mathbf{H}) P(\mathbf{W}) d\mathbf{W} + \ln P(\mathbf{H}) - \ln P(\mathbf{X}) \tag{1.51}$$

Now an optimal feature matrix $\mathbf{H}$ can be derived by maximizing Eq. (1.51) w.r.t. the entries of $\mathbf{H}$. To do so, we will utilize *Laplace's method*, which, being a saddle point method, is a common technique [26] to approximate high-dimensional integrals.

Laplace's Method

Let θ be an L-dimensional vector and $\tilde{P}(\theta)$ an unnormalized probability density. Expanding the logarithm of the integrand in

$$\int \tilde{P}(\theta) d\theta \tag{1.52}$$

by a Taylor series around its maximum θ^* yields

$$\ln \tilde{P}(\theta) \approx \ln \tilde{P}(\theta^*) - \frac{1}{2}(\theta - \theta^*)^T \mathbf{A}(\theta - \theta^*) + \cdots \tag{1.53}$$

where the first derivatives are zero due to the maximum condition and

$$A_{nm} = - \frac{\partial^2}{\partial\theta_n \partial\theta_m} \ln \tilde{P}(\theta) \bigg|_{\theta=\theta^*}$$

is the negative $L \times L$ Hessian matrix collecting the second order partial derivatives w.r.t. the entries of vector θ.

Exponentiating both sides of Eq. (1.53) yields

$$\tilde{P}(\theta) \approx \tilde{P}(\theta^*) \exp\left(-\frac{1}{2}(\theta - \theta^*)^T \mathbf{A}(\theta - \theta^*)\right) \tag{1.54}$$

and the original integral can be approximated by the well-known L-dimensional Gaussian

$$\int \tilde{P}(\theta) d\theta \approx \tilde{P}(\theta^*) \sqrt{\frac{(2\pi)^L}{\det(\mathbf{A})}} \tag{1.55}$$

Thus, using Laplace's method, the integral in Eq. (1.50) can be approximated by a Gaussian integral around the maximal argument $\mathbf{W}^*$ of its integrand, yielding

$$P(\mathbf{X}|\mathbf{H}) = \int P(\mathbf{X}|\mathbf{W},\mathbf{H})P(\mathbf{W})d\mathbf{W} \approx P(\mathbf{X}|\mathbf{W}^*,\mathbf{H})P(\mathbf{W}^*)(2\pi)^{\frac{NK}{2}}(\det(\mathbf{A}(\mathbf{W}^*)))^{-\frac{1}{2}} \tag{1.56}$$

where

$$\mathbf{A}(\mathbf{W}^*) =: -\nabla_W\nabla_W\{\ln[P(\mathbf{X}|\mathbf{W},\mathbf{H})P(\mathbf{W})]\}|_{\mathbf{W}=\mathbf{W}^*} \tag{1.57}$$

is a $(NK \times NK)$-dimensional matrix containing the second-order partial derivatives w.r.t. all entries W_{nk}, and ∇_W denotes the Nabla operator.

Utilizing this approximation and reordering terms, Eq. (1.51) simplifies to

$$\ln P(\mathbf{H}|\mathbf{X}) \approx \ln P(\mathbf{X}|\mathbf{W}^*,\mathbf{H}) + \ln P(\mathbf{W}^*) + \ln P(\mathbf{H}) - \frac{1}{2}\ln\det(\mathbf{A}(\mathbf{W}^*)) + \frac{NK}{2}\ln(2\pi) - \ln P(\mathbf{X}) \tag{1.58}$$

The first term in Eq. (1.58) represents the log-likelihood, where $\mathbf{W}^*$ denotes the matrix $\mathbf{W}$ which maximizes the sum $\ln P(\mathbf{X}|\mathbf{W},\mathbf{H}) + \ln P(\mathbf{W})$. The second and third terms represent prior distributions of the latent variables $\mathbf{W}$ and $\mathbf{H}$. While the two last terms are constants w.r.t. $\mathbf{W}$ and $\mathbf{H}$, the fourth term contains the logarithm of the determinant of a Hessian matrix w.r.t. the elements of $\mathbf{W}$. This latter term will be evaluated next whereby, remember, we will restrict ourselves to the case of a Gaussian likelihood function and sufficiently flat prior distributions.

1.3.2.2 The Bayesian Optimality Condition

Let the likelihood be of the form (1.45),

$$P(\mathbf{X}|\mathbf{W},\mathbf{H}) = \prod_n\prod_m \frac{1}{\sqrt{2\pi}\sigma_r}\exp\left(-\frac{1}{2}\left(\frac{X_{nm} - [\mathbf{WH}]_{nm}}{\sigma_r}\right)^2\right) \tag{1.59}$$

Then, the first-order partial derivatives of $\ln P(\mathbf{X}|\mathbf{W},\mathbf{H})$ are given by

$$\frac{\partial}{\partial W_{ab}}\ln(P(\mathbf{X}|\mathbf{W},\mathbf{H})) = \frac{1}{\sigma_r^2}\sum_m (X_{am} - [WH]_{am})H_{bm} \tag{1.60}$$

and the second-order derivatives are

$$\frac{\partial^2}{\partial W_{ab}\partial W_{cd}}\ln(P(\mathbf{X}|\mathbf{W},\mathbf{H})) = -\frac{1}{\sigma_r^2}\sum_m H_{bm}H_{dm}\delta_{ac} \tag{1.61}$$

Further, we assume the prior $P(\mathbf{W})$ to be sufficiently flat so that the second-order derivatives $\nabla_W \nabla_W \ln P(\mathbf{W})$ vanish and the resulting $NK \times NK$-dimensional Hessian simplifies to

$$\mathbf{A}(\mathbf{W}^*) = -\nabla_W \nabla_W \ln P(\mathbf{X}|\mathbf{W}, \mathbf{H})|_{\mathbf{W}=\mathbf{W}^*} \tag{1.62}$$

and exhibits a simple block structure as it is built of N^2 copies of the matrix $\mathbf{HH}^T$:

$$\mathbf{A} = -\frac{1}{\sigma_r^2} \left(\begin{array}{ccc|c|ccc} [HH^T]_{11} & 0 & 0 & \cdots & [HH^T]_{1K} & 0 & 0 \\ 0 & \ddots & 0 & \cdots & 0 & \ddots & 0 \\ 0 & 0 & [HH^T]_{11} & \cdots & 0 & 0 & [HH^T]_{1K} \\ \hline [HH^T]_{12} & 0 & 0 & \cdots & [HH^T]_{2K} & 0 & 0 \\ 0 & \ddots & 0 & \cdots & 0 & \ddots & 0 \\ 0 & 0 & [HH^T]_{12} & \cdots & 0 & 0 & [HH^T]_{2K} \\ \hline & \vdots & & \ddots & & \vdots & \\ \hline [HH^T]_{1K} & 0 & 0 & \cdots & [HH^T]_{KK} & 0 & 0 \\ 0 & \ddots & 0 & \cdots & 0 & \ddots & 0 \\ 0 & 0 & [HH^T]_{1K} & \cdots & 0 & 0 & [HH^T]_{KK} \end{array} \right)$$

For simplicity, we rescale the Hessian matrix according to $\tilde{\mathbf{A}} := \sigma_r^2 \mathbf{A}$. Now having to determine the determinant of the Hessian, we proceed by employing well-known results from linear algebra, i.e., the determinant of a matrix is the product of its eigenvalues and the eigenvalues of a triangular matrix are the diagonal elements. Indeed, the $(NK \times NK)$-dimensional Hessian matrix $\tilde{\mathbf{A}}$ can be transformed into a triangular matrix by using exactly the same row manipulations which are necessary to convert the $(K \times K)$-dimensional matrix $\mathbf{HH}^T$ into triangular form. Moreover, if $\lambda_1, \ldots, \lambda_K$ denote the eigenvalues of $\mathbf{HH}^T$, then the Hessian $\tilde{\mathbf{A}}$ possesses eigenvalues $(\lambda_1)^N, \ldots, (\lambda_K)^N$ and its determinant is obtained as

$$\det(\tilde{\mathbf{A}}) = \prod_{k=1}^{K} (\lambda_k)^N = \left(\prod_{k=1}^{K} \lambda_k \right)^N = (\det(\mathbf{HH}^T))^N \tag{1.63}$$

Reverting the rescaling, we finally have

$$\det(\mathbf{A}) = \det(\sigma_r^{-2} \tilde{\mathbf{A}}) = \sigma_r^{-2NK} \det(\tilde{\mathbf{A}}) = \sigma_r^{-2NK} (\det(\mathbf{HH}^T))^N \tag{1.64}$$

Summarizing the deductions above, an optimal feature matrix **H** can be obtained from maximizing

$$\ln P(\mathbf{H}|\mathbf{X}) \approx -\frac{1}{2\sigma_r^2}\sum_n\sum_m \left(X_{nm} - [\mathbf{W}^*\mathbf{H}]_{nm}\right)^2 - \frac{N}{2}\ln\det(\mathbf{H}\mathbf{H}^T) + \ln P(\mathbf{W}^*) + \ln P(\mathbf{H}) + const. \tag{1.65}$$

w.r.t. **H**, where $const. = \frac{NK}{2}\ln(2\pi) + NK\ln\sigma_r - \ln P(\mathbf{X})$ collects all terms which do not depend on the latent variables.

Note that $P(\mathbf{W}^*)$ was assumed representing a flat prior with nonzero values only on $\mathbb{R}_+$. If, in addition, we also choose $P(\mathbf{H})$ sufficiently flat and recall the definition of $\mathbf{W}^*$ from the saddle point approximation, the maximization of Eq. (1.65) w.r.t. **H** becomes the minimization, w.r.t. **W** and **H**, of the following cost function

$$D_{BOC}(\mathbf{X}, \mathbf{W}\mathbf{H}) := \frac{1}{2}\sum_n\sum_m (X_{nm} - [\mathbf{W}\mathbf{H}]_{nm})^2 + \frac{N\sigma_r^2}{2}\ln\det(\mathbf{H}\mathbf{H}^T) \tag{1.66}$$

The latter represents a Euclidean cost function for NMF with an additional volume constraint and closely resembles the cost function deduced within our geometric approach to NMF formalized in the algorithm *detNMF* (see Sect. 1.3.1.1).

In summary, in the absence of specific knowledge on the factor matrices, i.e., expressing our ignorance by flat priors of the latent variables, an optimal solution to a given NMF problem, with known model complexity K and Gaussian error statistics, is obtained by minimizing the Euclidean norm of the reconstruction error and constraining the optimization process by enforcing a minimal volume spanned by the extracted feature vectors. The trade-off between the reconstruction error and the volume constraint depends on the number of data vectors N and the spread parameter σ_r reflecting noise contributions.

1.3.3 Algorithms for Minimum Volume NMF

1.3.3.1 detNMF

In [3, 11] we introduced a new NMF algorithm named *detNMF* which directly implements a determinant criterion. As an objective function, we chose the regularized least squares cost function

$$D_{\det}(\mathbf{X}, \mathbf{W}\mathbf{H}) = \sum_{n=1}^{N}\sum_{m=1}^{M}(X_{nm} - [\mathbf{W}\mathbf{H}]_{nm})^2 + \lambda\det(\mathbf{H}\mathbf{H}^T), \tag{1.67}$$

where $\lambda > 0$ is a, usually small, positive regularization parameter. Optimization of the constrained cost function (1.67) can be achieved via gradient descent techniques according to

$$\frac{\partial D_{\text{det}}}{\partial H_{km}} = 2\sum_n \sum_m (X_{nm} - [\mathbf{WH}]_{nm}) \cdot \frac{\partial}{\partial H_{km}} \left(-\sum_{k'} W_{nk'} H_{k'm'} \right) + \lambda \frac{\partial}{\partial H_{km}} \det(\mathbf{HH}^T)$$
$$= -2\sum_n (X_{nm} - [\mathbf{WH}]_{nm})\, W_{nk} + \lambda \frac{\partial}{\partial H_{km}} \det(\mathbf{HH}^T). \tag{1.68}$$

The regularizing term $\det(\mathbf{HH}^T)$ is differentiable, and its partial derivatives are given by

$$\frac{\partial \det(\mathbf{HH}^T)}{\partial H_{km}} = 2\det(\mathbf{HH}^T)[(\mathbf{HH}^T)^{-1}\mathbf{H}]_{km} = 2\det(\mathbf{HH}^T)[\mathbf{H}^{\#}]_{km} \tag{1.69}$$

where $\mathbf{H}^{\#}$ represents the *Moore–Penrose pseudoinverse* of $\mathbf{H}$. Thus the following update rule results:

$$\begin{aligned} H_{km} &\leftarrow H_{km} - \eta^H_{km} \frac{\partial}{\partial H_{km}} E(\mathbf{X}, \mathbf{WH}) \\ &= H_{km} + 2\eta^H_{km} \left(\sum_n (X_{nm} - [\mathbf{WH}]_{nm})\, W_{nk} + \lambda \frac{\partial}{\partial H_{km}} \det(\mathbf{HH}^T) \right) \\ &= H_{km} + 2\eta^H_{km} \left([\mathbf{W}^T\mathbf{X} - \mathbf{W}^T\mathbf{WH}]_{km} - \lambda \det(\mathbf{HH}^T) \left[\mathbf{H}^{\#}\right]_{km} \right) \end{aligned} \tag{1.70}$$

However, such update rules are delicate because of the need for a proper choice of the learning rate η. But considering the fixed point equation

$$\frac{\partial}{\partial H_{km}} E(\mathbf{X}, \mathbf{WH}) = 0 \Rightarrow \frac{[\mathbf{W}^T\mathbf{X}]_{km} - \lambda \det(\mathbf{HH}^T) \left[\mathbf{H}^{\#}\right]_{km}}{[\mathbf{W}^T\mathbf{WH}]_{km}} = 1 \tag{1.71}$$

the additive learning rule (1.70) can be converted to a multiplicative update rule by setting

$$\eta^H_{km} = \frac{H_{km}}{2(\mathbf{W}^T\mathbf{WH})_{km}} \tag{1.72}$$

Due to the absence of additional constraints on the weights, the canonical updates for the W_{nk} can be used. Finally, the following multiplicative update rules (1.73)–(1.76) are obtained within the algorithm *detNMF*:

update:

$$H_{km} \leftarrow H_{km} \cdot \frac{[\mathbf{W}^T\mathbf{X}]_{km} - \lambda \det(\mathbf{HH}^T)[\mathbf{H}^{\#}]_{km}}{[\mathbf{W}^T\mathbf{WH}]_{km}} \tag{1.73}$$

$$W_{nk} \leftarrow W_{nk} \cdot \frac{[\mathbf{XH}^T]_{nk}}{[\mathbf{WHH}^T]_{nk}} \tag{1.74}$$

normalize:

$$H_{km} \leftarrow \frac{H_{km}}{\sqrt{\sum_m (H_{km})^2}} \tag{1.75}$$

$$W_{nk} \leftarrow W_{nk} \cdot \sqrt{\sum_m (H_{km})^2} \tag{1.76}$$

These update rules resemble the well-known Lee–Seung learning rules for NMF [8, 15] to which they reduce in the limit $\lambda \to 0$. In the latter case, the reconstruction error is guaranteed not to increase under the updates [15, 29, 43]. Since the algorithm is initialized by random matrices $\mathbf{W}$ and $\mathbf{H}$, a general rule for an optimal choice of the parameter λ is hard to find. However, if λ is kept small enough throughout, thus keeping the impact of the penalty term onto the optimization of the cost function moderate, then the reconstruction error did not increase during any iteration, and very satisfactory results were obtained with the above algorithm.

Historically, the algorithm *detNMF* was developed by our group in 2007 and published in a conference proceedings in 2009 [11]. A similar concept of a volume constraint for NMF has been independently developed to solve spectral unmixing problems by NMF [34]. In hyperspectral unmixing applications, the idea of an enclosing simplex with minimal volume is known as *endmember extraction* [35]. A later study found both, Miao's [34] and our approach [11] comparably useful as an input for a Bayesian sampling procedure, noting that our approach is simpler [44].

In all cases mentioned, the minimum volume constraint was motivated by heuristic considerations. Note that the volume constraint derived in the *BOC* (Eq. 1.66) contains an additional logarithm. In [45], finally, $\ln|\det(\mathbf{H})|$ for a square matrix $\mathbf{H}$ is used.

1.3.3.2 Multilayer NMF

The determinant criterion can also be related to a practice called multilayer technique. As first mentioned in [46] and studied in more detail in [47], a sequential decomposition of non-negative matrices as a cascade of unmixing systems usually yields a better performance than one single decomposition. Using the determinant criterion, this effect can intuitively be explained as follows:

In the first step of the multilayer approach, the decomposition $\mathbf{X} \approx \mathbf{W}^{(1)}\mathbf{H}^{(1)}$ is performed where $\mathbf{W}^{(1)}$ is a $N \times K$ and $\mathbf{H}^{(1)}$ is a $K \times M$-matrix. The second step then considers the decomposition: $\mathbf{W}^{(1)} \approx \mathbf{W}^{(2)}\mathbf{H}^{(2)}$. The lth step then uses the result from the previous $l-1$ step and decomposes $\mathbf{W}^{(l-1)} \approx \mathbf{W}^{(l)}\mathbf{H}^{(l)}$ where $\mathbf{H}^{(l)}$ is a square $K \times K$-matrix. Thus, a L-stage multilayer system performs the decomposition

$$\mathbf{X} \approx \mathbf{W}^{(L)}\mathbf{H}^{(L)}\mathbf{H}^{(L-1)} \ldots \mathbf{H}^{(2)}\mathbf{H}^{(1)}. \tag{1.77}$$

Setting $\mathbf{H}^{(L)}\mathbf{H}^{(L-1)}\dots\mathbf{H}^{(2)}\mathbf{H}^{(1)} =: \mathbf{H}$, the determinant criterion reads

$$\begin{aligned}\det\left(\mathbf{H}\mathbf{H}^T\right) &= \det\left(\mathbf{H}^{(L)}\dots\mathbf{H}^{(2)}\mathbf{H}^{(1)}\mathbf{H}^{(1)T}\mathbf{H}^{(2)T}\dots\mathbf{H}^{(L)T}\right)\\ &= \det\left(\mathbf{H}^{(L)}\mathbf{H}^{(L)T}\dots\mathbf{H}^{(2)}\mathbf{H}^{(2)T}\mathbf{H}^{(1)}\mathbf{H}^{(1)T}\right)\\ &= \det\left(\mathbf{H}^{(L)}\mathbf{H}^{(L)T}\right)\dots\det\left(\mathbf{H}^{(2)}\mathbf{H}^{(2)T}\right)\det\left(\mathbf{H}^{(1)}\mathbf{H}^{(1)T}\right). \quad (1.78)\end{aligned}$$

Resolving the scaling indeterminacy in every subproblem by normalizing the rows of $\mathbf{H}^{(l)}$ such that $\sqrt{\sum_j (H_{kj}^{(l)})^2} = 1 \; \forall \; k = 1, \dots, K$ implies that

$$\det\left(\mathbf{H}^{(l)}\mathbf{H}^{(l)T}\right) \leq 1 \quad (1.79)$$

for every term $l = 1, \dots, L$ in the last line of Eq. (1.78).

As discussed above, the lth subsystem $\mathbf{W}^{(l-1)} \approx \mathbf{W}^{(l)}\mathbf{H}^{(l)}$ can be interpreted such that $\mathbf{W}^{(l)}$ are the coefficients when representing the data $\mathbf{W}^{(l-1)}$ in a basis $\mathbf{H}^{(l)}$. Any solution positions the basis vectors $\mathbf{H}_{k*}^{(l)}$ on the periphery of the data cloud $\mathbf{W}^{(l-1)}$. According to the determinant criterion, the solution comprising the smallest volume is optimal for the particular subproblem. The more demixing sub-steps are performed, the higher the probability is to obtain a solution with a small determinant in at least one sub-step, and, accordingly, overall a smaller determinant results.

1.3.4 Bayesian Model Order Selection

Until now, it was assumed that the actual number K of intrinsic modes was known at the outset. The following discussion thus deals with the model order selection problem of how to automatically obtain an appropriate model order K, representing the number of inherent modes in the data. Thus we finally discuss a second level of Bayesian inference where a variational approximation to the log-evidence $\ln P(\mathbf{X}|K)$ is derived by integrating over all latent variables. We show how to derive a *Variational Bayes NMF Algorithm* (VBNMF) which allows an iterative and approximative computation of the evidence $P(\mathbf{X}|K)$ the data provide about model $\mathcal{M}_K$. Variational methods generally intend to transform a complicated integration problem such as the computation of the data evidence into an optimization problem. We show how Jensen's inequality can be employed to convert the, generally intractable, logarithm of an expectation, such as the log-evidence, into expectations of logarithms with respect to properly chosen auxiliary distributions $Q(\mathbf{W}, \mathbf{H})$. Employing the latter induces a lower bound to the log-evidence which turns into an identity if the auxiliary distributions equal the posterior distributions of the parameters $Q(\mathbf{W}, \mathbf{H}) = P(\mathbf{W}, \mathbf{H}|\mathbf{X})$. The latter affords the computation of a high-dimensional integral which, in practice, is substituted by either Monte Carlo sampling procedures or a variational approach.

We follow the latter philosophy and show how, by employing Jensen's inequality a second time, a *Variational Bayes NMF* algorithm can be obtained in case of a Gaussian data likelihood and rectified Gaussian prior distributions of the parameters W_{ik}, H_{kj}. Thus, learning proper parameters H_{kj}, W_{ik} affords a number of simplifying assumptions which, as a byproduct, relieve the model order selection problem and allow to determine an optimal number of modes K from the data automatically. Finally we discuss the relation of the resulting update rules to the famous Lee–Seung multiplicative update rules and show how the latter result from our VBNMF algorithm by deriving a MAP version of it, i.e., replacing expectations by the modes of the corresponding distributions.

A Bayesian approach to estimating the complexity of an NMF model is closely related to automatic relevance detection (ARD) schemes which have been mentioned first by MacKay [48] and have been put forward by Bishop [49] in a *Bayesian PCA* approach. In NMF settings, Cemgil [39] discusses the Poisson likelihood case, while Schmidt [50] demonstrates Gibbs sampling schemes in connection with Chib's method for the Gaussian likelihood case. Furthermore, Zhong and Girolami [51] and Schmidt and Mørup [52] use two different reversible jump MCMC approaches for model order selection in the Gaussian likelihood case. The latter group [44] also discusses a Bayesian NMF approach with a volume prior and an inference scheme based on Gibbs sampling. In [40], the use of Markov Chain Monte Carlo (MCMC) and variational Bayes methods is sketched to perform Bayesian inference on NMF problems using the Frobenius norm, the KL divergence and the Itakura–Saito divergence as cost functions. Variational methods have been extensively used in the field of Bayesian learning [53–58]. A variational Bayes approach for NMF using Poisson statistics has first been proposed by A. Cemgil [39]. Bayesian approaches to model order selection for NMF have further been discussed in [59] in a MAP estimation framework and in relation with β-divergences [60].

Associating the complexity of an NMF model with the number K of intrinsic modes, a second level of Bayesian inference can be employed for comparing alternative models with each other by evaluating the log-evidence for model $\mathscr{M}_K$

$$\ln P(\mathbf{X}|K) = \ln \left\{ \int \int P(\mathbf{X}|\mathbf{W},\mathbf{H}) P(\mathbf{W},\mathbf{H}|K) d\mathbf{W} d\mathbf{H} \right\} \quad (1.80)$$

while integrating over all H_{km} as well as over all W_{nk} for different values of K. However, computing the latter distribution is not feasible, in general, due to the high-dimensional integral involved which is intractable. One way around this difficulty is provided by sampling procedures as mentioned in [39, 50] where Gibbs sampling is applied to draw samples from the true posterior distribution $P(\mathbf{W},\mathbf{H}|\mathbf{X})$. The approach pursued here rather forms a variant of the *Variational Bayesian EM Algorithm* proposed in [55] and further discussed in [56, 61].

Rendering Eq. (1.80) tractable consists in applying Jensen's inequality [62], which brings the logarithm inside the integral:

$$\begin{aligned}\ln P(\mathbf{X}|K) &\geq \int\int Q(\mathbf{W},\mathbf{H}) \ln \frac{P(\mathbf{X}|\mathbf{W},\mathbf{H})P(\mathbf{W},\mathbf{H})}{Q(\mathbf{W},\mathbf{H})} d\mathbf{W}d\mathbf{H} \\ &= \langle \ln P(\mathbf{X}|\mathbf{W},\mathbf{H})\rangle_{Q(\mathbf{W},\mathbf{H})} + \left\langle \ln \frac{P(\mathbf{W},\mathbf{H})}{Q(\mathbf{W},\mathbf{H})} \right\rangle_{Q(\mathbf{W},\mathbf{H})} \\ &:= \mathscr{B}_Q \end{aligned} \tag{1.81}$$

This step induces a lower bound $\mathscr{B}_Q$ on $\ln P(\mathbf{X}|K)$ where $Q(\mathbf{W},\mathbf{H})$ is an auxiliary variational distribution obeying the normalization constraint $\int Q(\mathbf{W},\mathbf{H})d\mathbf{W}d\mathbf{H} = 1$, and where $\langle\cdot\rangle_Q$ denotes the expectation w.r.t. the distribution Q. It can be shown [61, 63] that equality holds if the variational distribution equals the posterior distribution of the latent variables.

In practice, we choose Q of some simple form, such that the expectations $\langle\ldots\rangle_Q$ can be computed. Thus, in the following we restrict ourselves to the Gaussian likelihood for NMF (Eq. 1.59) and, furthermore, assume completely factorizable distributions

$$Q(\mathbf{W},\mathbf{H}) = \prod_n\prod_k Q(W_{nk})\prod_k\prod_m Q(H_{km}) := Q(\mathbf{W})Q(\mathbf{H}) \tag{1.82}$$

$$P(\mathbf{W},\mathbf{H}) = \prod_n\prod_k P(W_{nk})\prod_k\prod_m P(H_{km}). \tag{1.83}$$

These assumptions render the lower bound $\mathscr{B}_Q$ to the log-evidence of the form

$$\begin{aligned}\mathscr{B}_Q = &-\frac{NM}{2}\ln(2\pi) - NM\ln\sigma_r \\ &-\frac{1}{2\sigma_r^2}\sum_n\sum_m\Big\{X_{nm}^2 - 2X_{nm}\sum_k \langle W_{nk}\rangle_{Q(W_{nk})}\langle H_{km}\rangle_{Q(H_{km})} \\ &\qquad + \left\langle([\mathbf{W}\mathbf{H}]_{nm})^2\right\rangle_{Q(\mathbf{W}),Q(\mathbf{H})}\Big\} \\ &+\sum_n\sum_k\left\langle \ln\frac{P(W_{nk})}{Q(W_{nk})}\right\rangle_{Q(W_{nk})} + \sum_k\sum_m\left\langle \ln\frac{P(H_{km})}{Q(H_{km})}\right\rangle_{Q(H_{km})}\end{aligned} \tag{1.84}$$

Next we simplify the computation of the joint ensemble average $\langle\ldots\ldots\rangle_{Q(\mathbf{W}),Q(\mathbf{H})}$ by a decoupling approximation. To do so, we introduce additional parameters q_{nkm}, which follow the closure relation $\sum_k q_{nkm} = 1$ for every observation X_{nm}, such that

$$\left\langle([\mathbf{W}\mathbf{H}]_{nm})^2\right\rangle_{Q(\mathbf{W}),Q(\mathbf{H})} = \left\langle\left(\sum_k q_{nkm}\frac{W_{nk}H_{km}}{q_{nkm}}\right)^2\right\rangle_{Q(\mathbf{W}),Q(\mathbf{H})} \tag{1.85}$$

If we now apply Jensen's inequality once more, we obtain

$$\begin{aligned}\left\langle([\mathbf{WH}]_{nm})^2\right\rangle_{Q(\mathbf{W}),Q(\mathbf{H})} &\leq \sum_k q_{nkm}\left\langle\left(\frac{W_{nk}H_{km}}{q_{nkm}}\right)^2\right\rangle_{Q(\mathbf{W}),Q(\mathbf{H})} \\ &= \sum_k \frac{1}{q_{nkm}}\left\langle(W_{nk})^2\right\rangle_{Q(W_{nk})}\left\langle(H_{km})^2\right\rangle_{Q(H_{km})} \end{aligned} \tag{1.86}$$

With this approximation, a new lower bound $\mathscr{B}_q$ results according to

$$\begin{aligned}\mathscr{B}_Q \geq \mathscr{B}_q :=& -\frac{NM}{2}\ln(2\pi) - NM\ln\sigma_r \\ &-\frac{1}{2\sigma_r^2}\sum_n\sum_m\Bigg\{X_{nm}^2 - 2X_{nm}\sum_k\langle W_{nk}\rangle_{Q(W_{nk})}\langle H_{km}\rangle_{Q(H_{km})} \\ &\qquad + \sum_k\frac{1}{q_{nkm}}\left\langle(W_{nk})^2\right\rangle_{Q(W_{nk})}\left\langle(H_{km})^2\right\rangle_{Q(H_{km})}\Bigg\} \\ &+\sum_n\sum_k\left\langle\ln\frac{P(W_{nk})}{Q(W_{nk})}\right\rangle_{Q(W_{nk})} + \sum_k\sum_m\left\langle\ln\frac{P(H_{km})}{Q(H_{km})}\right\rangle_{Q(H_{km})}\end{aligned} \tag{1.87}$$

Now $\mathscr{B}_q$ in Eq. (1.87) contains only one-dimensional expectations and always forms a lower bound to the log-evidence $\ln P(\mathbf{X}|\mathscr{M}_K)$.

Given these approximations, an optimal log-evidence $\ln P(\mathbf{X}|K)$ can be obtained by maximizing $\mathscr{B}_q$ w.r.t. all auxiliary variational distributions $Q(W_{nk})$ and $Q(H_{km})$ as well as the decoupling parameters q_{nkm} and the noise parameter σ_r. Hence, we are now ready to derive an algorithm which maximizes $\mathscr{B}_q$. Thereby the following steps need to be performed:

1. First, optimal variational densities Q can be obtained by solving the fixed point equation

$$\frac{\partial}{\partial Q(H_{km})}\left(\mathscr{B}_q - \lambda_{km}\left(\int Q(H_{km})dQ(H_{km}) - 1\right)\right) \stackrel{!}{=} 0 \tag{1.88}$$

   where λ_{km} is a Lagrange parameter ensuring that $Q(H_{km})$ is a density (see, e.g., [36] for a brief introduction to variational calculus). Straightforward computation leads to

$$Q(H_{km}) \propto P(H_{km})\exp\left(\alpha_{km}^H(H_{km})^2 + \beta_{km}^H H_{km}\right) \tag{1.89}$$

   where

$$\alpha_{km}^H = -\frac{1}{2\sigma_r^2}\sum_n\frac{1}{q_{nkm}}\left\langle(W_{nk})^2\right\rangle_{Q(W_{nk})} \tag{1.90}$$

$$\beta_{km}^H = \frac{1}{\sigma_r^2}\sum_n X_{nm}\langle W_{nk}\rangle_{Q(W_{nk})} \tag{1.91}$$

Note that this relation holds whatever the priors $P(H_{km})$ might be. Corresponding optimal densities $Q(W_{nk})$ can be computed in complete analogy to the calculation given above.

2. The optimal decoupling parameters q_{nkm} can be deduced from a corresponding fixed point equation (see Eq. (1.88))

$$q_{nkm} = \frac{\sqrt{\left\langle (W_{nk})^2 \right\rangle_{Q(W_{nk})} \left\langle (H_{km})^2 \right\rangle_{Q(H_{km})}}}{\sum_l \sqrt{\left\langle (W_{il})^2 \right\rangle_{Q(W_{il})} \left\langle (H_{lj})^2 \right\rangle_{Q(H_{lj})}}} \tag{1.92}$$

3. Next, the optimal noise parameter σ_r is obtained as

$$\sigma_r^2 = \frac{1}{NM} \sum_n \sum_m \left\{ X_{nm}^2 - 2X_{nm} \sum_k \langle W_{nk} \rangle_{Q(W_{nk})} \langle H_{km} \rangle_{Q(H_{km})} + \sum_k \frac{1}{q_{nkm}} \left\langle (W_{nk})^2 \right\rangle_{Q(W_{nk})} \left\langle (H_{km})^2 \right\rangle_{Q(H_{km})} \right\} \tag{1.93}$$

4. Computing the expectations $\langle .... \rangle_Q$ affords choosing proper priors for the latent variables **W** and **H**. Remember that we need priors which are nonzero only for non-negative real-valued arguments. Such priors are available via truncated Gaussian distributions [57]. They derive from a Gaussian distribution by truncating all negative entries and re-normalizing its integral to unity. The resulting *truncated Gaussian distribution* is given by

$$\mathcal{N}^+(\theta|\mu, \sigma) =: \begin{cases} \frac{1}{Z_{\mathcal{N}^+}} \exp\left(-\frac{1}{2} \left(\frac{\theta - \mu}{\sigma} \right)^2 \right), & 0 \leq \theta < \infty \\ 0 & \text{otherwise} \end{cases} \tag{1.94}$$

The normalizing constant is given by

$$Z_{\mathcal{N}^+} = \frac{1}{2} \sigma \sqrt{2\pi} \, \mathrm{erfc} \left(-\frac{\mu}{\sigma\sqrt{2}} \right) \quad \text{with} \quad \mathrm{erfc}(z) =: \frac{2}{\sqrt{\pi}} \int_z^\infty \exp(-t^2) dt \tag{1.95}$$

where $\mathrm{erfc}(z)$ represents the complementary error function. The expectations required for our purpose are then given by

$$\langle \theta \rangle_{\mathcal{N}^+} = \mu + \frac{\sigma^2}{Z_{\mathcal{N}^+}} \exp\left(-\frac{\mu^2}{2\sigma^2} \right) \tag{1.96}$$

$$\langle \theta^2 \rangle_{\mathcal{N}^+} = \sigma^2 + \mu \langle \theta \rangle_{\mathcal{N}^+} \tag{1.97}$$

1.3.4.1 The VBNMF Update Rules

If, in the following, we choose rectified Gaussian prior distributions for the latent variables according to

$$\begin{aligned} P(W_{nk}) &= \mathscr{N}^+(W_{nk}|\mu_{W_{0k}}, \sigma_{W_{0k}}) \\ P(H_{km}) &= \mathscr{N}^+(H_{km}|\mu_{H_{k0}}, \sigma_{H_{k0}}) \end{aligned} \tag{1.98}$$

and if one such distribution is assumed for each column $\mathbf{W}_{*k}$ or each row $\mathbf{H}_{k*}$, we are expressing our "belief" that the latent variables for each intrinsic mode k are drawn from the same *a priori* distribution. Considering the principle of conjugate priors, we deduce from Eq. (1.89) corresponding auxiliary variational distributions as

$$Q(W_{nk}) = \mathscr{N}^+(W_{nk}|\mu_{W_{nk}}, \sigma_{W_{nk}})$$

and

$$Q(H_{km}) = \mathscr{N}^+(H_{km}|\mu_{H_{km}}, \sigma_{H_{km}}) \tag{1.99}$$

where

$$\mu_{W_{nk}} = \frac{\mu_{W_{0k}} \sigma_{W_{k0}}^{-2} + (\sigma_r)^{-2} \sum_m X_{nm} \langle H_{km} \rangle}{\sigma_{W_{0k}}^{-2} + (\sigma_r)^{-2} \sum_m q_{nkm}^{-1} \langle H_{km}^2 \rangle} \tag{1.100}$$

$$\sigma_{W_{nk}} = \left(\sigma_{W_{0k}}^{-2} - (\sigma_r)^{-2} \sum_m q_{nkm}^{-1} \langle H_{km}^2 \rangle \right)^{-\frac{1}{2}} \tag{1.101}$$

and

$$\mu_{H_{km}} = \frac{\mu_{H_{k0}} \sigma_{H_{k0}}^{-2} + (\sigma_r)^{-2} \sum_n X_{nm} \langle W_{nk} \rangle}{\sigma_{H_{k0}}^{-2} + (\sigma_r)^{-2} \sum_n q_{nkm}^{-1} \langle W_{nk}^2 \rangle} \tag{1.102}$$

$$\sigma_{H_{km}} = \left(\sigma_{H_{k0}}^{-2} - (\sigma_r)^{-2} \sum_n q_{nkm}^{-1} \left\langle W_{nk}^2 \right\rangle \right)^{-\frac{1}{2}} \tag{1.103}$$

If we have no knowledge on the *actual* values of the hyperparameters $\mu_{W_{0k}}, \sigma_{W_{0k}}$ and $\mu_{H_{k0}}, \sigma_{H_{k0}}$, they can be updated by iteratively solving the corresponding fixed point equations:

$$\frac{\partial \mathscr{B}_q}{\partial \mu_{W_{0k}}} \stackrel{!}{=} 0 \qquad \frac{\partial \mathscr{B}_q}{\partial \sigma_{W_{0k}}} \stackrel{!}{=} 0 \tag{1.104}$$

$$\frac{\partial \mathscr{B}_q}{\partial \mu_{H_{k0}}} \stackrel{!}{=} 0 \qquad \frac{\partial \mathscr{B}_q}{\partial \sigma_{H_{k0}}} \stackrel{!}{=} 0 \tag{1.105}$$

Finally, the overall *Variational Bayes NMF* (VBNMF) algorithm, as proposed above, can be summarized by

- set hyperparameters $\mu_{H_{k0}}, \sigma_{H_{k0}}, \mu_{W_{0k}}, \sigma_{W_{0k}}$
- initialize $\mu_{H_{km}}, \sigma_{H_{km}}, \mu_{W_{nk}}, \sigma_{W_{nk}}$ and σ_r
- Repeat
  - (1) compute all q_{nkm} → (Eq. 1.92)
  - (2) update all $\mu_{H_{km}}$ and $\sigma_{H_{km}}$ → (Eqs. (1.102) and (1.103))
  - (3) update all $\mu_{W_{nk}}$ and $\sigma_{W_{nk}}$ → (Eqs. (1.99) and (1.101))
  - (4) update σ_r → (Eq. 1.93)
  - (5) update hyperparameters → (Eqs. (1.104) and (1.105))
  - (6) compute $\mathscr{B}_q$ → (Eq. 1.87)

  until convergence

Steps (4) and (5) are optional if no prior knowledge exists. Also step (6) can be performed only occasionally testing the state of convergence of the algorithm. Since the latter represents a generalized Expectation–Maximization (EM) algorithm [55, 56, 61], convergence is assured.

Some final remarks are in order here:

- Notice that a strict Bayesian approach would require assigning prior distributions to the parameters σ_r as well and integrating them out. However, following the discussion in [64, 65], we assume that the distribution $P(\sigma_r|\mathbf{X}, \mathbf{W}, \mathbf{H})$ is sharply peaked at the maximum, and maximize $\mathscr{B}_q$ w.r.t. σ_r instead of performing an additional integration.
- Further notice that the quadratic terms in Eq. (1.84) can be expressed in terms of first- and second-order expectations. Hence, the second application of Jensen's inequality is not a necessary requirement to construct a variational Bayes algorithm for Gaussian likelihood NMF.

For example, the approach presented in [40] relies on the property that Gaussian and Poisson distributions are closed under summation and report EM and variational Bayes NMF algorithms concluding that the resulting update rules differ from Lee and Seung's multiplicative update equations [15] in the Euclidean case, while they are identical in the KL case. Our method differs from the one presented in [40] mainly in utilizing the second Jensen bound from Eqs. (1.84)–(1.87). In the following paragraph we will show that our algorithm indeed reduces to the famous Lee–Seung multiplicative update equations.

1.3.4.2 Relation to the Lee–Seung Method

It is interesting to note that the VBNMF algorithm derived above is a straightforward Bayesian generalization of the multiplicative NMF algorithm for the Euclidean distance proposed by [15]. To see this, we derive a MAP version of the VBNMF algorithm as follows:

- Replacing the expectations of the form $\langle W_{nk}\rangle_{Q(W_{nk})}$, $\langle H_{km}\rangle_{Q(H_{km})}$ by the MAP values W^*_{nk} H^*_{km}, the update equation for the decoupling parameters q_{nkm} Eq. (1.92) reduces to

$$q_{nkm} = \frac{W^*_{nk} H^*_{km}}{\sum_l W^*_{il} H^*_{lj}} \tag{1.106}$$

- Rectified Gaussian priors $P(H_{km})$ and $P(W_{nk})$ become flat priors if the spread parameters $\sigma_{H_{k0}}$ and $\sigma_{W_{0k}}$ grow large. Then, in the limit $\sigma_{H_{k0}}, \sigma_{W_{00}} \to \infty$, the first terms in the numerator and denominator of Eqs. (1.102) and (1.99) tend toward zero and the update rules for the mean parameters $\mu_{H_{km}}$ and $\mu_{W_{nk}}$ reduce to

$$\mu_{H_{km}} = \frac{\sum_n X_{nm} W^*_{nk}}{\sum_n \frac{(W^*_{nk})^2}{q_{nkm}}} \quad \text{and} \quad \mu_{W_{nk}} = \frac{\sum_m X_{nm} H^*_{km}}{\sum_m \frac{(H^*_{km})^2}{q_{nkm}}} \tag{1.107}$$

- Plugging in the q_{nkm} from Eq. (1.106) finally yields

$$\begin{aligned} \mu_{H_{km}} &= \frac{\sum_n X_{nm} W^*_{nk}}{\sum_n \frac{(W^*_{nk})^2 \sum_l W^*_{il} H^*_{lj}}{W^*_{nk} H^*_{km}}} \\ &= H^*_{km} \frac{\sum_n X_{nm} W^*_{nk}}{\sum_n W^*_{nk} \sum_l W^*_{il} H^*_{lj}} \\ &= H^*_{km} \frac{[\mathbf{W^*}^T \mathbf{X}]_{km}}{[\mathbf{W^*}^T \mathbf{W^*} \mathbf{H^*}]_{km}} \end{aligned} \tag{1.108}$$

and similarly

$$\mu_{W_{nk}} = W^*_{nk} \frac{[\mathbf{X} \mathbf{H^*}^T]_{nk}}{[\mathbf{W} \mathbf{H^*} \mathbf{H^*}^T]_{nk}} \tag{1.109}$$

The last expressions correspond exactly to the update equations proposed by Lee and Seung [15].

1.3.5 VBNMF Simulations on Toy Data Sets

Artificial toy data sets $\mathbf{X}$ were generated according to Eq. (1.3) assuming a Gaussian noise distribution and a rectified Gaussian prior distribution of the weights $\mathbf{W}$. The simulations employed the VBNMF algorithm and the parameters were chosen as follows:

- The elements of the error term were drawn from a Gaussian distribution with zero mean and variance σ_r^2.
- Feature vectors were predefined forming the following matrix

$$\mathbf{H} = \begin{pmatrix} 3 & 1 & 0 & 1 & 3 \\ 1 & 2 & 3 & 4 & 5 \\ 0 & 3 & 3 & 3 & 0 \end{pmatrix} \tag{1.110}$$

- The rows of the corresponding weight matrix **W** were generated from rectified Gaussian distributions with parameters set to $\mu_{0k} = 3k$, $\sigma_{0k} = 3k$.
- The model order varied over the range $K = 2, \ldots, 8$.
- The hyperparameters were updated algorithmically, since in general their distribution is not available as prior knowledge.

The simulations tested several instances of the noise parameter σ_r. Occasionally occurring negative data entries, induced by the noise, were set to zero. Strictly speaking, this truncation constitutes a still modest violation of the likelihood assumption which, however, becomes more severe with an increasing noise level. In the following, we discuss two prototypical cases where

1. matrix **W** obeys the prior assumptions and is generated from a rectified Gaussian distribution (Fig. 1.9, left)
2. matrix **W** is drawn from a mixture of a rectified Gaussian and a peak at zero (Fig. 1.10, left)

Simulation results, presented in Figs. 1.9 and 1.10, indicate that

- if the assumption concerning the distribution of **W** holds true, i.e.,

$$P(W_{nk}) = \mathcal{N}^+ \tag{1.111}$$

the correct number of features is, for various noise levels, reliably detected by the VBNMF algorithm (see Fig. 1.9) as indicated through a maximum in the $\mathcal{B}_q$ plot.
- Next the assumption on the prior of **W** has been violated deliberately by adding a delta peak at the origin of the rectified Gaussian distribution (Fig. 1.10, left) according to

$$P(W_{nk}) = \tau \mathcal{N}^+ + (1 - \tau)\delta(W_{nk}) \tag{1.112}$$

Such priors could represent sparse priors in practical situations. In the noise-free case, the true number $K = 3$ of internal modes is detected (see Fig. 1.10). If the noise level is low ($\sigma_r = 1$ here), the lower bound $\mathcal{B}_q$ of the true log-evidence still identifies the correct order though it also favors a slightly larger model order, i.e., it overestimates the number of features slightly. But in case of high noise levels, the method fails completely and bad estimates of the number of inherent degrees of freedom result.

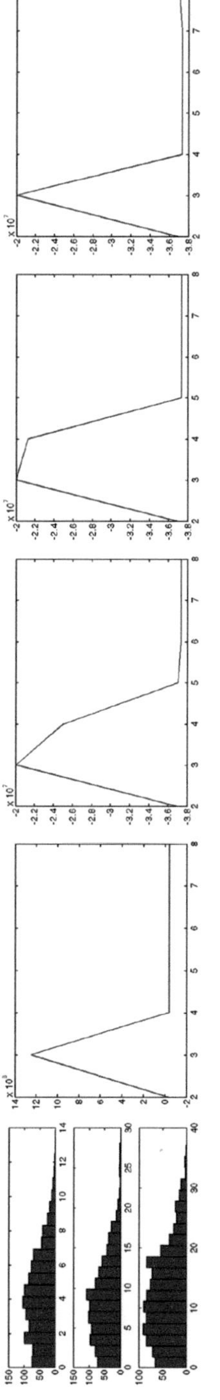

Fig. 1.9 *From left to right* Histograms of the original columns of the 1000×3-dimensional matrix $\mathbf{W}$ drawn from a rectified Gaussian $P(W_{nk}) = \mathcal{N}^+$, followed by the log-evidence bound $\mathcal{B}_q$ w.r.t. the number of components K for various noise levels $\sigma_r = 0$, $\sigma_r = 1$, $\sigma_r = 10$, $\sigma_r = 20$. The correct number of sources $K = 3$ is recognized in all cases

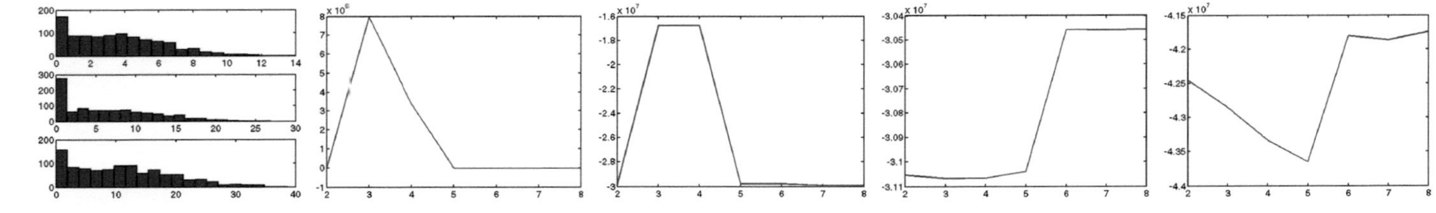

Fig. 1.10 *From left to right* Histograms of the original columns of the 1000×3-dimensional matrix $\mathbf{W}$ drawn from a rectified Gaussian with an additional peak at zero, i.e., $P(W_{nk}) = \tau\mathcal{N}^{+} + (1-\tau)\delta(W_{nk} - 0)$. The log-evidence bound $\mathscr{B}_q$ w.r.t. the number of components K for various reconstruction errors $\sigma_r = 0$, $\sigma_r = 1$, $\sigma_r = 10$, $\sigma_r = 20$. Results refer to a situation where the assumption about the prior distribution is violated because of the additional density at the origin. Only at low noise levels the true number of sources $K = 3$ is recognized

1.3.5.1 *VBNMF* for Real-World Data

The performance of the *VBNMF* algorithm is tested also on a real-world data set collected during manufacturing 1000 semiconductor wafers [1]. A plot of the lower bound $\mathscr{B}_q$ to the log-evidence $\ln P(\mathbf{X}|K)$ exhibits a maximum at $K = 5$, indicating five intrinsic failure patterns, which underly the observed wafer maps (see Fig. 1.11).

The *VBNMF* algorithm explains the observations $\mathbf{X}$ with non-negative factors $\mathbf{W}$ and $\mathbf{H}$ with concomitant rectified Gaussian priors. Figure 1.12 compares maximum likelihood solutions $\mathbf{W}^{ALS}$, $\mathbf{H}^{ALS}$, obtained with the alternating least squares (ALS) algorithm, with *VBNMF* results $\langle\mathbf{W}\rangle_Q$, $\langle\mathbf{H}\rangle_Q$ for $K = 15$. While the ALS algorithm uses all 15 components $\mathbf{H}_{k*}$ to explain the data, the VBNMF only retains five components and renders all remaining components irrelevant. The histograms of the weights $\langle\mathbf{W}_{*k}\rangle_Q$ show that weight distributions corresponding to irrelevant components resemble delta distributions, while the ones related with the intrinsic failure patterns correspond to rectified Gaussians. Note that this behavior resembles automatic relevance determination (ARD) [59], although ARD has not been applied explicitly here.

All ALS maximum likelihood counterparts (Fig. 1.12, bottom left) show a clear peak at the origin of the distribution of the weights $\mathbf{W}_{*k}$. This suggests a more flexible shape for the prior distributions, such as a superposition of a rectified Gaussian and a delta peak at the origin

$$P(W_{ik}) = \tau\mathscr{N}^+ + (1-\tau)\delta(W_{ik}). \tag{1.113}$$

The latter represents a limiting case of a Gaussian mixture model.

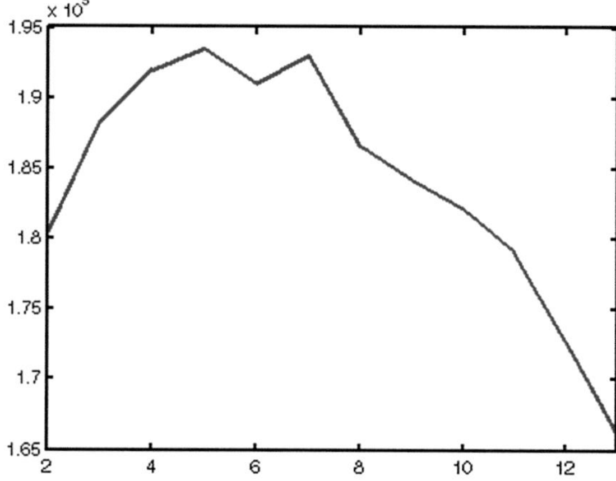

Fig. 1.11 The lower bound $\mathscr{B}_q$ of the log-evidence as function of the number of elementary components K of the real-world data set exhibits a maximum at $K = 5$

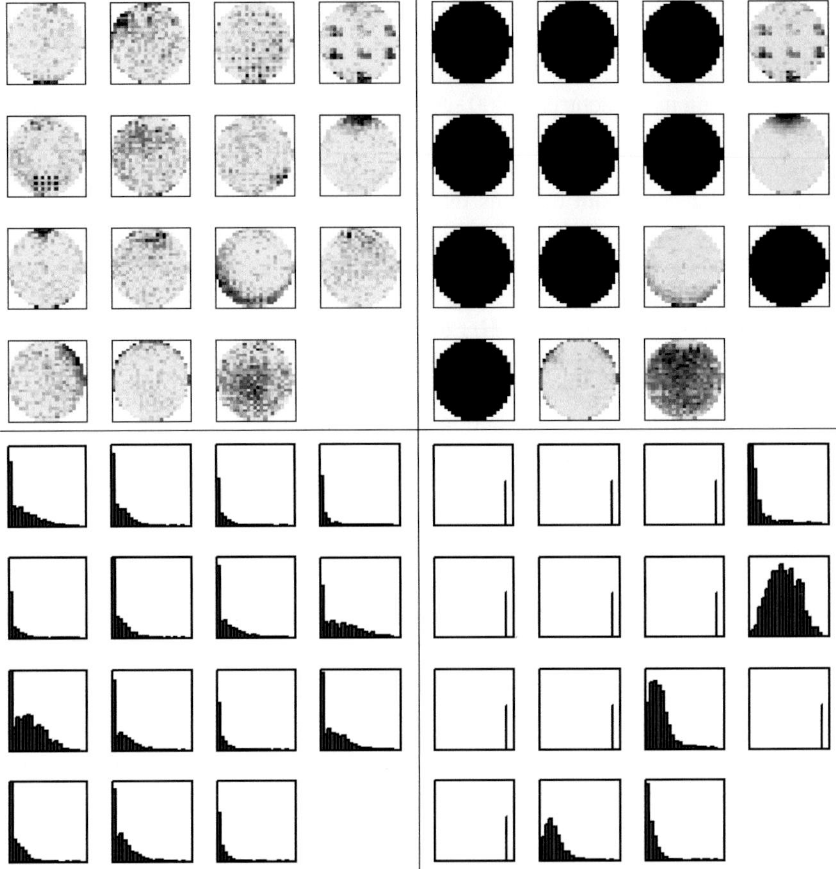

Fig. 1.12 Automatic relevance determination by *VBNMF* shown in an real data example where $K = 15$ *top, left* basic patterns $\mathbf{H}^{ALS}$ gained by approximate maximum likelihood *top, right* expected basic patterns $\langle \mathbf{H} \rangle_Q$ gained by *VBNMF*, initialized by $\mathbf{H}^{ALS}$. *Bottom, left* histograms of weights $\mathbf{W}^{ALS}$ *bottom, right* histograms of expected weights $\langle \mathbf{W} \rangle_Q$ Only relevant components are retained while the others are switched off by setting the respective basic pattern $\mathbf{H}_{k*}$ to a constant value and the distribution of the related weights $\mathbf{W}_{*k}$ to a narrow Gaussian peak

1.4 Conclusion

The manuscript summarized our work on probabilistic NMF ranging from *binNMF* via *detNMF* to VBNMF. All algorithms were illustrated through applications to wafer maps, i.e., binary failure patterns on silicon wafers emerging during microchip fabrication. The basic idea is to consider such failure patterns as superpositions of a finite number of intrinsic failure patterns inherent to and characteristic of the manufacturing process chain. Exploratory matrix factorization techniques, more specifically non-negative matrix factorization (NMF), has been employed to blindly decompose observed wafer maps into such underlying patterns. A probabilistic treatment has been applied throughout.

Binary NMF *BinNMF* was presented utilizing an extension of NMF to binary data sets. An optimization technique was presented which maximizes a Bernoulli likelihood. This is achieved by preprocessing a fast alternating least squares algorithm on a quadratic approximation to the true likelihood function followed by gradient ascent refinement. The performance of the overall procedure was demonstrated on an artificial toy data example and real-world data sets. The model was placed into a context with existing literature on related topics noting that none of the common techniques is suitable to find a satisfactory solution to the described problem.

The studies on binary data sets have been pursued further by employing Bernoulli statistics and a factorization of the related matrix of log-odds. The resulting *logistic NMF* model puts a constraint onto one of the factor matrices of the log-odds decomposition, i.e., the matrix of basis vectors $\mathbf{W}$ or $\tilde{\mathbf{W}}$, to render the estimated basis system strictly non-negative or even binary, while the other factor, i.e., the matrix $\mathbf{H}$ of related features, may take on any value in $\mathbb{R}$. Thus adding up strictly non-negative basis vectors may end up at points outside of the positive orthant occasionally. Hence, the main difference between logistic PCA, *binNMF* and the proposed *logistic NMF* model rests in the additional constraints imposed onto the models. Logistic PCA does not constrain the log-odds parameters any further. The nonlinear *binNMF* model asks for strictly non-negative entries to both factor matrices similar to the linear aspect Bernoulli model, which constrains both factors to the range [0, 1] but employs a linear decomposition only. The *logistic NMF* model proposed here only requires a set of strictly non-negative basis vectors but allows for negative feature components resulting, occasionally, in a subtractive superposition of the underlying basis vectors. Thereby the proposed *logistic NMF* model places itself in between logistic PCA and *binNMF* as the corresponding limiting cases. A strictly non-negative set of basis vectors is appropriate for many real-world problems while corresponding feature vectors may have negative components allowing for some subtractive superpositions of the basis vectors to represent any observations considered. An application of the method to the USPS data set reveals the performance of the various variants of the model and shows good reconstruction quality even with a low rank binary basis set.

Determinant NMF *DetNMF* introduced a determinant criterion thus constraining the possible solutions of an exact NMF problem. Geometrically, this criterion formulates a minimum volume constraint on the subspace spanned by the basis vectors and emphasizes unique best solutions for a given problem. *DetNMF* was used in illustrative toy examples which represent two extreme data distributions. In these extremal settings, the *detNMF* algorithm was contrasted with a sparse NMF variant to demonstrate that sparseness constraints can be a misleading restriction while the determinant criterion is a more general approach. Moreover, the determinant criterion provides a very concrete explanation why a multilayer technique usually improves the performance of any NMF algorithm.

Finally, with *VBNMF* we addressed problems concerning NMF decompositions $\mathbf{X} \approx \mathbf{WH}$, such as uniqueness and model order selection, by employing Bayesian methods of various sophistication. Considering Gaussian error statistics of an L_2-

norm reconstruction error, and a flat prior on the feature distribution, we derived a Bayesian optimality criterion (BOC) for an optimal solution, given the correct factorization rank K. This BOC is achieved by integrating over all possible weight matrices $\mathbf{W}$ to estimate the posterior distribution of the feature matrix $\mathbf{H}$. To perform the integration, we utilized a Laplace approximation of the related integral. In case of a Gaussian likelihood this yields a simple criterion which is identical to the heuristical *detNMF* criterion [11]. Next, we introduced a novel variational Bayes algorithm which solves the problem of model order selection and implements an automatic relevance detection mechanism which is able to discover the actual number of components underlying artificially constructed toy data sets. The resulting VBNMF algorithm was also shown to be a Bayesian generalization of the famous Lee–Seung NMF algorithm. Closing the circle, we explored the potential of the VBNMF algorithm to analyze binary data sets. There the underlying assumptions relating to the likelihood function and the prior distributions are only roughly valid. Still very acceptable results are obtained.

Acknowledgments Support by Infineon Technologies AG and the DAAD (PPP Luso-Alemã, PPP Hispano-Alemana) is gratefully acknowledged.

References

1. R. Schachtner, G. Pöppel, E.W. Lang, IEEE Trans. Circuits Syst. I **57**(7), 1439 (2010)
2. A.M. Tomé, R. Schachtner, V. Vigneron, C.G. Puntonet, E.W. Lang, Multidimensional Systems and Signal Processing (2013), pp. 1–19. doi:10.1007/s11045-013-0240-9
3. R. Schachtner, G. Pöppel, E.W. Lang, Digit. Signal Process. **21**(4), 528 (2011)
4. R. Schachtner, G. Pöppel, A. Tomé, C. Puntonet, E.W. Lang, Neurocomputing **138**, 142 (2014)
5. A. Cichocki, arXiv:1403.2048v4 [cs.ET] 24 Aug 2014 (2014), pp. 1–30
6. P. Paatero, U. Tapper, Environmetrics **5**(2), 111 (1994)
7. P. Paatero, Chemom. Intell. Lab. Syst. **37**, 23 (1997)
8. D.D. Lee, H.S. Seung, Nature **401**(6755), 788 (1999)
9. A. Cichocki, S. Amari, R. Zdunek, A.H. Phan, *Non-negative Matrix and Tensor Factorizations: Applications to Exploratory Multi-way Data Analysis and Blind Source Separation* (Wiley-Blackwell, Oxford, 2009)
10. R. Schachtner, G. Pöppel, E.W. Lang, in *32ndAnnual Meeting of the German Classification Society (GfKl)* (2008), pp. 755–764
11. R. Schachtner, G. Pöppel, A.M. Tomé, E.W. Lang, in *Proceedings of 8th International Conference on Independent Component Analysis and Signal Separation, ICA 2009*, Paraty, Brazil, 15–18 March 2009, pp. 106–113. doi:10.1007/978-3-642-00599-2_14
12. R. Schachtner, G. Pöppel, E.W. Lang, in *Proceedings of the 2nd International Workshop on Cognitive Information Processing on Elba Island, CIP2010* (2010), pp. 57–62
13. R. Schachtner, G. Pöppel, A.M. Tomé, E.W. Lang, Pattern Recognit. Lett. **45**, 251 (2014). doi:10.1016/j.patrec.2014.04.013
14. A. Cichocki, R. Zdunek, *Advances in Neural Networks (ISNN 2007)*. Lecture Notes in Computer Science, vol. 4493 (Springer, Berlin, 2007), pp. 793–802
15. D.D. Lee, H.S. Seung, NIPS **13**, 556–562 (2001)
16. A.I. Schein, L.K. Saul, L.H. Ungar, in *Proceedings of the 9th International Workshop on Artificial Intelligence and Statistics* (2003), pp. 1–8
17. S. Lee, J.Z. Huang, J. Hu, Ann. Appl. Stat. **4**(3), 1579 (2010)

18. E. Meeds, Z. Ghahramani, R.M. Neal, S.T. Roweis, Bernoulli **19**(8), 977 (2007)
19. E. Lang, R. Schachtner, D. Lutter, D. Herold, A. Kodewitz, F. Blöchl, F.J. Theis, I.R. Keck, J.G. Saez, P.G. Vilda, A.M. Tomé, New advances in biomedical signal processing, in *Exploratory Matrix Factorization Techniques for Large Scale Biomedical Data Sets*, ed. by J.M. Górriz-Sáez, E.W. Lang, J. Ramírez (Bentham Science Publishers, 2011), pp. 26–47. doi:10.2174/97816080521891110101
20. A. Cichocki, R. Zdunek, S. Amari, in *2006 IEEE International Conference on Acoustics Speech and Signal Processing, ICASSP 2006*, Toulouse, France, 14–19 May 2006 (2006), pp. 621–624. doi:10.1109/ICASSP.2006.1661352
21. M.E. Tipping, in *Proceedings of Advances in Neural Information Processing Systems II (NIPS 1998)* (1999), pp. 592–598
22. A. Kabán, E. Bingham, T. Hirsimäki, in *Proceedings of 4th SIAM International Conference on Data Mining* (2004), pp. 462–466
23. M. Collins, S. Dasgupta, R.E. Schapire, in *NIPS 11* (2001), pp. 592–598
24. E. Bingham, A. Kaban, M. Fortelius, Pattern Anal. Appl. **12**(1), 55 (2009)
25. K.H. Knuth, in *Proceedings of the 13th European Signal Processing Conference (EUSIPCO 2005)* (2005), pp. 1–8. arXiv:1311.3001 [stat.ML]
26. D.J.C. MacKay, *Information Theory, Inference, and Learning Algorithm* (Cambridge University Press, Cambridge, 2003). http://www.inference.phy.cam.ac.uk/mackay/itila/
27. A. Cichocki, R. Zdunek, S. Amari, *Csiszar's Divergences for Non-Negative Matrix Factorization, Family of New Algorithms* (Springer, Berlin, 2006), pp. 32–39
28. I. Dhillon, S. Sra, in *Proceedings of Neural Information Processing Systems (NIPS)* (2005)
29. S. Sra, I.S. Dhillon, Nonnegative matrix approximation: algorithms and applications. Technical report, Computer Sciences, University of Texas, Technical Report, # TR-06-27 (2006)
30. A. Cichocki, R. Zdunek, S. Amari, IEEE Signal Process. Mag. **142**, 142 (2008)
31. P.K. Hopke, in *EPA Workshop Proceedings, Materials from the Work shop on UNMIX and PMF as Applied to PM2.5.* (2000). http://www.epa.gov/ttnamti1/files/ambient/pm25/workshop/laymen.pdf
32. P. Sajda, S. Du, T. Brown, L. Parra, R. Stoyanova, in *4th International Symposium on Independent Component Analysis and Blind Signal Separation* (2003), pp. 71–76
33. P. Sajda, S. Du, T.R. Brown, R. Stoyanova, D.C. Shungu, X. Mao, L.C. Parra, IEEE Trans. Med. Imaging **23**(12), 1453 (2004)
34. L. Miao, H. Qi, IEEE Trans. Geosci. Remote Sens. **45**(3), 765 (2007)
35. M.D. Craig, IEEE Trans. Geosci. Remote Sens. **32**(3), 542 (1994)
36. C.M. Bishop, *Neural Networks for Pattern Recognition* (Oxford University Press, Oxford, 1996)
37. P. Hoyer, in *Proceedings of the IEEE Workshop on Neural Networks for Signal Processing* (2002), pp. 557–565
38. M.N. Schmidt, H. Laurberg, Comput. Intell. Neurosci. **1** (2008). doi:10.1155/2008/361705
39. A.T. Cemgil, Comput. Intell. Neurosci. **1** (2009). doi:10.1155/2009/785152
40. C. Févotte, A.T. Cemgil, in *Proceedings of 17th European Signal Processing Conference (EUSIPCO'09)* (2009), pp. 1–5
41. S. Moussaoui, D. Brie, O. Caspary, A. Mohammad-Djafari, in *Proceedings of IEEE International Conference on Acoustics, Speech, and Signal Processing* (2004), pp. 485–48
42. T.O. Virtanen, A.T. Cemgil, S.J. Godsill, in *Proceedings of IEEE ICASSP* (2008), pp. 1–4
43. C.J. Lin, IEEE Trans. Neural Netw. **18**(6), 1589 (2007)
44. M. Arngren, M.N. Schmidt, J. Larsen, J. Signal Process. Syst. **65**(3), 479 (2010)
45. D. Zhou, H.Y. Gao, Y.J. Zhang, Adv. Mater. Res. **651**, 858 (2013)
46. K. Stadlthanner, F. Theis, C. Puntonet, J.M. Górriz, A.M. Tomé, E.W. Lang, in *ISBMDA*. LNCS (LNBI), vol. 3745 (Springer, Heidelberg, 2005), pp. 137–148
47. A. Cichocki, R. Zdunek, Int. J. Neural Syst. **17**(6), 431 (2007)
48. D.J.C. MacKay, Ensemble learning and evidence maximization. Technical report, Cavendish Laboratory, University of Cambridge (1995)
49. C.M. Bishop, in *Advances in Neural Information Processing Systems NIPS* (1999), pp. 382–388

50. M.N. Schmidt, O. Winther, L.K. Hanse, in *International Conference on Independent Component Analysis and Signal Separation*. Lecture Notes in Computer Science (LNCS), vol. 5441 (Springer, New York, 2009), pp. 540–547
51. M. Zhong, M. Girolami, J. Mach. Learn. Res. **5**, 663 (2009)
52. M.N. Schmidt, M. Mørup, in *European Signal Processing Conference (EUSIPCO)* (2010)
53. M.I. Jordan, Z. Ghahramani, T.S. Jaakkola, L.K. Saul, Mach. Learn. **37**, 183 (1998)
54. C.M. Bishop, in *Proceedings 9-th International Conference on Artificial Neural Networks, ICANN* (1999), pp. 509–514
55. H. Attias, *Advances in Neural Information Processing Systems, NIPS 12* (MIT Press, Cambridge, 2000)
56. Z. Ghahramani, M.J. Beal, *Advances in Neural Information Processing Systems NIPS 13* (MIT Press, Cambridge, 2001), pp. 507–513
57. M. Harva, A. Kabán, Signal Process. **87**(3), 509 (2007)
58. A. Kabán, E. Bingham, Neurocomputing **71**(10–12), 2291 (2008)
59. V.Y.F. Tan, C. Fevotte, in *Proceedings of Workshop on Signal Processing with Adaptative Sparse Structured Representations (SPARS'09)* (2009), pp. 1–5
60. V.Y.F. Tan, C. Fevotte, IEEE Trans. Pattern Anal. Mach. Intell. **35**(7), 1592 (2013)
61. M.J. Beal, Z. Ghahramani, Bayesian Anal. **1**(4), 793 (2006)
62. J.L.W.V. Jensen, Acta Math. **30**, 175 (1906)
63. Z. Ghahramani, *Advanced Lectures on Machine Learning* (2004) pp. 72–112
64. D.J.C. MacKay, Neural Comput. **4**(3), 415 (1992)
65. D.J.C. MacKay, *Maximum Entropy and Bayesian Methods* (Kluwer Academic Publishers, Boston, 1996), pp. 43–60

Chapter 2
Nonnegative Matrix Factorizations for Intelligent Data Analysis

G. Casalino, N. Del Buono and C. Mencar

Abstract We discuss nonnegative matrix factorization (NMF) techniques from the point of view of intelligent data analysis (IDA), i.e., the intelligent application of human expertize and computational models for advanced data analysis. As IDA requires human involvement in the analysis process, the understandability of the results coming from computational models has a prominent importance. We therefore review the latest developments of NMF that try to fulfill the understandability requirement in several ways. We also describe a novel method to decompose data into user-defined—hence understandable—parts by means of a mask on the feature matrix, and show the method's effectiveness through some numerical examples.

Keywords Nonnegative matrix factorization · Intelligent data analysis

2.1 Introduction

The amount of available data has grown dramatically over the past 50 years. Every year more than 200 Exabytes of data are generated on Internet.[1] Huge quantities of digital data are produced daily from different sources: numerical data from satellites or sensors, textual data (both structured and unstructured) from websites, emails,

[1]Source: Cisco® Visual Networking Index (VNI) Forecast (2010–2015).

G. Casalino (✉) · C. Mencar
Department of Informatics University of Bari, University Campus
"Ernesto Quagliariello", Via E. Orabona, 4, 70125 Bari, Italy
e-mail: gabriella.casalino@uniba.it

C. Mencar
e-mail: corrado.mencar@uniba.it

N. Del Buono
Department of Mathematics University of Bari, University Campus
"Ernesto Quagliariello", Via E. Orabona, 4, 70125 Bari, Italy
e-mail: nicoletta.delbuono@uniba.it

G.R. Naik (ed.), *Non-negative Matrix Factorization Techniques*,
Signals and Communication Technology, DOI 10.1007/978-3-662-48331-2_2

forums, newsgroups, public and private digital archives, images, and videos, are just some examples. Data overload is a fact of life for all of us in the information era.

Although this profusion of information potentially allows to satisfy all information needs, it also presents some limits; the larger is the amount of data the fewer are the possibilities to capture, discover, and understand useful knowledge to guide action or decision making [8, 76].

Clearly human capabilities prove to be unsuitable to process big amounts of data, therefore automatic mechanisms, which are able to assist humans in extracting useful information and knowledge from rapidly growing volumes of digital data are indispensable and an extensive effort of research in this direction has been made in the last years.

Intelligent data analysis (IDA) aims to the intelligent application of human expertise and computational models for advanced data analysis. Automatic tools, which strive for involving the analyst in the process of data analysis and extracting useful patterns from big data, can be enumerated among IDA methods. In this scenario, techniques coming from different areas (such as statistics, artificial intelligence, data mining, machine learning, optimization, dynamic programming), which favor the interaction with users and produce understandable knowledge, could be favorably exploited in IDA.

Nonnegative matrix factorizations (NMF) are powerful techniques recently proposed to uncover latent low-dimensional structures intrinsic in high-dimensional data and provide a nonnegative, part-based, representation of data [5, 25, 40, 69, 70, 115]. Nonnegativity enhances meaningful interpretations of mined information and distinguishes NMF from other traditional dimensionality reduction algorithms, such as principal component analysis (PCA) [65] or singular value decomposition (SVD) [45].

However, the understandability of the results, obtained by applying classical NMF, is not guaranteed a priori, as they often do not correspond with the intuitive notions of parts in the original data. Several variants of constraints and various regularization terms have been proposed to improve NMF capabilities so as to make the extracted parts easier to understand by the data analyst.

This chapter aims to review such techniques from the point of view of IDA, by stressing on their understandability capabilities and usefulness as tools for IDA.

Along with the review of NMF techniques suited for IDA, we describe an approach for injecting user knowledge in the factorization process, by masking the factor matrix (one of the products of NMF) [13]. Masking enables the decomposition of data into user-defined parts, which are consequently easy to understand by the analyst. The results of Masked NMF enables the analyst to understand which subset of the available data are best represented by the specified parts, thus extracting potentially useful knowledge from large quantities of data.

In the next section, we give an overview of IDA and its objective, while we focus on NMF techniques in Sects. 2.3 and 2.4. In Sect. 2.4.3.1, we describe Masked NMF along with some numerical examples to show the effectiveness of the method. In Sect. 2.5, the use of the NMF algorithms for intelligently analyze educational data is illustrated. Future perspectives are sketched in the conclusive section.

2.2 Intelligent Data Analysis

Data are collections of values or measurements. They can be numbers, words, observations, or even descriptions of things. In this chapter, we will simply refer to data as a collection of numerical values recording the magnitude of different attributes and/or features that describe the problem under study.

Hand writes *"data analysis is what we do when we turn data into information"* [50]. Intelligent data analysis is the intelligent way to do it. Moreover, he gives a definition of information: *"It is what we extract from data when we attempt to answer some questions. Before extracting information which can shed light on a question, one must be clear about what that question is"*. This is a crucial point in IDA: the analysis is driven by the questions that the analyst wants to answer to, otherwise it would be *unintelligent*.

IDA is an iterative process that enables the combination of human expertise and computational models to automatically extract useful patterns, event correlations and in general, understandable knowledge which would otherwise remain hidden in the data under consideration [6]. A data analyst could be interested in describing data by finding patterns and anomalies, or just by summarizing them; in this case, the term *exploratory data analysis* is used. Contrariwise, when the analyst is interested in verifying some hypotheses about the structure in data, e.g., differences among groups of data, evolution of the attribute values, etc., the term *confirmatory data analysis* is used. IDA is a multidisciplinary discipline that comes from the intersection of several research fields, the most important ones are statistics and machine learning.

IDA and knowledge discovery from data (KDD) are tightly correlated, yet with some noteworthy differences. Both are aimed at identifying valid, novel, potentially useful, and ultimately understandable patterns in data [36]; however, IDA emphasizes the importance of the prior knowledge possessed by human experts that intelligently guide the analysis process in an interactive and iterative way [7]. Data mining is one step of the KDD process and refers to the set of tools that allow to *automatically* extract knowledge from large amounts of data [36]. However, a full automatization of the data analysis process is impossible [7], for this reason IDA is focused on the human contribution to the analysis process.

Holmes and Peek categorize IDA methods in three main classes: data exploration, classification and prediction, and dimensionality reduction [54]. Data exploration plays a fundamental role in data analysis. Analysts look at data for discovering relations among features, trends, anomalies or outliers, relations among features, classes, etc. Most of these techniques rely on visual tools to represent information. IDA-based approaches for data exploration integrate automatic techniques with a priori user knowledge in the exploration process, thus enabling user interaction. Classification and prediction methods are used in several domains dealing with real data. Machine learning literature provides many different techniques for classification (both supervised, semi-supervised, or unsupervised) and prediction. Most of them are based on some automatic learning tools to acquire knowledge that can be used for classifying (or predicting) unobserved data. However, only few of them are capable of yielding knowledge that is intelligible to users (e.g., knowledge expressed in form of rules), a mandatory requirement for their use within IDA. Learning interpretable knowledge from data is a topic of current research in Machine Learning and Computational Intelligence. In this context are located dimensionality reduction techniques that represent data in a reduced space through feature selection and extraction. This facilitates to manage, understand, and visualize data. Because of their tight relationship with NMF, a brief overview of such techniques is outlined in the following subsection.

2.2.1 Dimensionality Reduction Techniques

Often, in high-dimensional data not all the measured variables are "important" for understanding the underlying phenomena of interest. Hence, mechanisms that transform data and reduce the number of original variables are frequently used.

Let $X \in \mathbb{R}^{n \times m}$ be the observation data matrix, where each columns vector is composed by n observations for each of the m-dimensional variable in $x = (\mathbf{x_1}, \ldots, \mathbf{x_m})^\top$. In this formalization, the dimension of data is meant the number of variables that are measured on each observation, while the term dimensionality of X indicates the number m of original features. A dimensionality reduction method is a transformation of a given data matrix X into a meaningful representation $S \in \mathbb{R}^{n \times k}$ of reduced dimensionality $k \leq m$ [108]. The low-dimensional vectors $s = (\mathbf{s_1}, \ldots, \mathbf{s_k})^\top$, with $k \leq m$ capture information in the original data, according to some particular criteria. The components of s are called "hidden components" or "latent factors," while—depending on the particular research context one is working with— the m multivariate vectors are alternatively named "variables," "attributes" or "features." Dimensionality reduction methods mitigate *the curse of dimensionality* [1], which refers to difficulties related to data analysis when data dimensionality increases; these methods are able to overcome problems coming from data sparseness and noise, and can be adopted as a visualization tool to show multivariate data in a human intelligible form.

Dimensionality reduction techniques can be categorized in two classes: (i) *feature selection* and (ii) *feature extraction.* A feature selection method is a process that selects a subset of k original (and supposed relevant) features for spanning a reduced space that may better describe the phenomena of interest. Feature selection mechanisms reduce the computational costs, but a good trade-off between accuracy of the results and efficiency is needed.

On the other hand, feature extraction methods try to capture hidden properties of data and discover the minimum number of uncorrelated or lowly correlated factors that can be used to better describe the phenomena of interest. It is accomplished by the creation of new features obtained as functions of the original data. Reduction of the computational complexity of data both in time (for elaboration) and in space (for storage) and the discovery of latent structure hidden in data, (meaningful structures and/or unexpected relationships among variables) are some of the advantages resulting from feature extraction methods.

The simplest dimensionality reduction methods are linear and derive each of the $k \leq m$ components of the new variables in S as a linear combination of the original variables:

$$S = XA, \tag{2.1}$$

or equivalently

$$X = SB, \tag{2.2}$$

being $A \in \mathbb{R}^{m \times k}$ and $B \in \mathbb{R}^{k \times m}$ appropriate linear transformation weight matrices. Equation (2.2) makes clear the motivation why the new variables in S are called hidden or latent factors. (PCA) [56, 65, 93], factor analysis (FA) [102], independent component analysis (ICA) [60], linear discriminant analysis (LDA) [38], and CUR decomposition [85] are all well-known linear dimensionality reduction techniques used for analyzing multivariate data. Among linear dimensionality reduction methods, the most widely used in the context of IDA is PCA.

2.2.1.1 Principal Component Analysis

Principal component analysis is the best, in the least square error sense, linear dimensionality reduction technique [62, 65]. It is based on the covariance matrix of the variables and seeks to reduce the dimensionality of data matrix X by finding few orthogonal linear combinations (the principal components—PCs) of the original variables with the largest variance. The first PC is the linear combination of the original data with the largest variance; the second PC is the linear combination with the second largest variance and orthogonal to the first PC, and so on. The principal components are given by

$$Y = XU, \tag{2.3}$$

where $U \in \mathbb{R}^{m \times m}$ is an orthogonal weight matrix computed as the orthogonal factor of the spectral decomposition of the covariance matrix $X^\top X$ of the standardized data matrix X.[2] Therefore, the columns of the matrix U are the eigenvectors of the covariance matrix. These eigenvectors (principal axes) map a data vector from the original space of m variables to a new space of k variables which are uncorrelated over the dataset. Hence keeping only the first $k < m$ principal components a dimensional reduction on k-dimensional subspace of the original data is derived.

Moreover, it is proven that the transformed data matrix, obtained by only considering the first $k < m$ principal components, is the best least squares k-approximation of the original data X (this result is known as the *Eckart–Young–Mirsky theorem* [46]).

[2] Since the values of the variance of data depends on the scale of the variables, usually the original data contained in X are subject to a standardization process so that each variable has mean zero and standard deviation one.

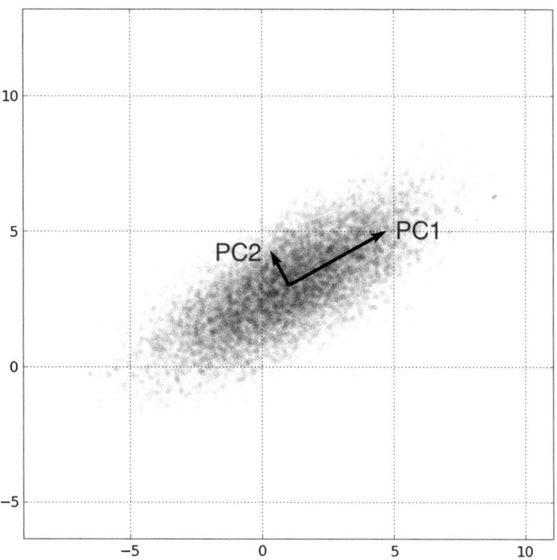

Fig. 2.1 Graphical illustration of PCA. From Wikipedia, the free encyclopedia

Figure 2.1 shows the behavior of PCA of a data matrix collecting points that belong to a bivariate Gaussian distribution centered in the coordinates $(1, 3)$. Standard deviation of data is 3 in the direction (0.878, 0.478) and 1 in the orthogonal direction. The first principal component ($PC1$) captures information in the direction of the maximum variability in data; instead, the second principal component ($PC2$) is orthogonal to the first one and captures information in the second most variable direction. The principal axes are therefore the bases of the rotated space and are centered in the center of the points. This is a simple example where the dimensionality of the original data space and that of the transformed one are the same. As an example of dimensionality reduction, one can represent the same points using the first principal axis only; in this case, a one-dimensional space is obtained where data points are projected onto.

In many applications, the most of data variance can be captured by the first two (or three) PCs; this makes the PCA a widely used visualization tool in IDA. However, even though the PCs are uncorrelated variables constructed as linear combinations (with mixed signs) of the original variables, and have some desirable properties (they are orthogonal and ordered in a decreasing manner w.r.t. the variance of original data), they do not necessarily correspond to meaningful physical quantities. Hence, a clear interpretation of the results provided by PCA is sometimes difficult to be derived.

To clarify this point consider the computer vision problem of human face recognition, where PCA has been largely adopted to obtain a set of basis images—*the eigenfaces*—that can be linearly combined to reconstruct images in the original dataset of face [107]. As it can be observed by Fig. 2.3 (left panel), eigenfaces are not physically intuitive and far to correspond to what humans use to explain why a face is a face. In particular, because of the presence of negative signs in the components of principal axes, PCA reconstructs the original data adding up some basis images and subtracting others; this may not make sense in some applications. A simply question can be posed:"What does it mean to subtract a face basis?"

These considerations can be extended to documents, genes, preferences, questionaries and to all nonnegative data. In the following Sect. 2.3, a review of Nonnegative Matrix Factorization is given. It is able to represent original data by only additive, not subtractive, combinations of some basis vectors. This characteristic of parts-based representation is appealing because it reflects the intuitive notion of combining parts to form a whole providing more distinct and clearer dimensionality reduction results and a easier understandability of the obtained results.

2.3 Nonnegative Matrix Factorization

Nonnegative matrix factorization (NMF) is a computational technique for linear dimensionality reduction of a given data matrix X, which is able to explain data in terms of additive combination of nonnegative factors that represent realistic *building blocks* for the original data (provided that data are nonnegative too) [25, 40, 69, 70, 115].

The nonnegativity constraint is useful for learning part-based representations and has a twofold motivation. First in many applications one knows that the quantities involved cannot be negative (for example by the rules of physics). Second, intuitively parts are generally combined additively (and not subtracted) to form a whole and physiological principles assume that humans learn objects as part-based [70]. Hence, nonnegativity potentially enhances meaningful interpretations of information mined from a given data matrix, allowing to a better understanding of the results obtained by the analysis process; this makes NMF a suitable computational models for IDA.

2.3.1 NMF Mathematical Formulation

Formally, given a nonnegative data matrix $X = [\mathbf{x}_1, \mathbf{x}_2, \ldots, \mathbf{x}_m] \in \mathbb{R}_+^{n \times m}$, where $\mathbf{x}_i \in \mathbb{R}_+^n$ are n-dimensional column vectors representing samples,[3] NMF aims to approximate X into the product of two lower rank nonnegative matrices—a *basis matrix* $W = [\mathbf{w}_1, \mathbf{w}_2, \ldots, \mathbf{w}_k] \in \mathbb{R}_+^{n \times k}$ and an *encoding matrix* $H = (h_{ij}) \in \mathbb{R}_+^{k \times m}$—such that

$$X \approx WH, \tag{2.4}$$

[3]Henceforth a matrix is denoted with an uppercase letter, e.g., X, its elements with the corresponding lowercase letter, e.g., x_{ij}, a column vector in lowercase boldface, e.g., $\mathbf{x}_i$.

or, equivalently,

$$\mathbf{x}_j \approx \sum_{i=1}^{k} \mathbf{w}_i h_{ij}. \tag{2.5}$$

where W and H both have nonnegative elements (namely, $W \geq 0$ and $H \geq 0$) and the product matrix (WH) is of rank k with $(n + m)k \leq nm$.

To compute a nonnegative matrix factorization (2.4) of a given data matrix X, some quality measures have to be taken into account to evaluate how well the product (WH) approximates the data matrix X. Particularly, some divergence function

$$D : \mathbb{R}_+^{n \times r} \times \mathbb{R}_+^{r \times m} \to \mathbb{R}_+,$$

can be adopted. It should be observed that the divergence D is a function of the factor matrices W and H, but it is also parametrized by the input data matrix X. This dependence can be expressed by writing $D(X; W, H)$ [104]. Using the previous formalization, the NMF problem may be rewritten as a nonlinear constrained optimization problem over the divergence D, that is,

$$\min_{W \geq 0, H \geq 0} D(X; W, H). \tag{2.6}$$

The most frequently adopted instance of (2.6) leads to the minimization of

$$\min_{W \geq 0, H \geq 0} D(X; W, H) = \|X - WH\|_F^2, \tag{2.7}$$

where $\|\cdot\|_F$ denotes the Frobenius norm. Many other divergence measures have also been used, the interested reader can refer to [30].

The NMF performs a conical coordinate transformation; indeed, geometrically the basis vectors generate a simplicial cone which contains the original data and which is contained in the positive orthant [24, 33, 59].

It should be pointed out that the value of the parameter k (the *rank* of the factorization) is problem dependent and user-specified. It identifies the number of factors to be used to explain data and plays a fundamental role in the factorization process. In fact, different values of k lead to different factorization results.

2.3.2 Interpretation of the Basis and Encoding Matrices

The results of NMF applied to a data matrix X have an immediate geometrical interpretation. According to Eq. (2.5), the columns of the matrix W are basis vectors spanning a subspace in $k \leq n$ dimensions, called NMF-subspace, while each column of the encoding matrix H represents the new coordinates of the corresponding data sample in the NMF-subspace. From a numerical point of view, each data sample is approximated by a linear combination of vectors in W, where the linear coefficients are grouped in a column of H corresponding to the data sample. Therefore, the elements h_{ij} codify the amount of the factors (i.e., the columns of W) used to reconstruct each sample of X in the NMF-subspace.

The coefficients h_{ij} in each column of H define the importance of each basis vector in approximating the data sample; if a coefficient is very small, then the corresponding basis vector is useless in approximating the sample. Under some hypotheses,[4] the basis vectors can be interpreted as prototypes of data clusters. In this case, the coefficients h_{ij} can be easily interpreted as membership degrees of each sample to each cluster.

Examples of successful applications of NMF are: basic student skills describing student questionnaire results in educational data mining [27]; topics represented as bag of words in text mining [14, 101]; anatomic parts of images describing human faces in face identification problems [49, 103]; part-based representation of digital characters for object recognition [48, 80]; community categories extracted to describe users networks [110]; diversified portfolio describing trends in stock markets in financial data mining [34, 99]; topics used to clusterize social tags data [18]; users–items relations in recommender systems [47]; chemical constituents in air pollution revelations [55, 66]; musical instrument frequencies for music classification [2–4]; endmembers of constituent materials of hyperspectral images [10, 42, 44, 63, 84, 92]; genes in microarray data [9, 12, 29, 32, 67, 86]. Other successful applications of NMF, where interpretability is a key requirement, belong to molecular pattern discovery [39, 67] and object detection [15].

A key aspect of NMF, that is advantageous for its application in IDA, stands in the possibility of approximating data samples as linear combination of factors, where the factors are subsets of the same features used to represent data samples. Therefore, unlike other low-rank approximation techniques, NMF allows to represent data as composition of parts, being each part expressed with the same features used in data. This makes the results of NMF easily interpretable for the analyst, who can intelligently guide the factorization process, in order to achieve results that are interesting and useful for understanding the problem at hand.

2.3.3 Comparison of NMF and PCA

As stated before, PCA can be used as a tool in IDA because of its dimensionality reduction and visualization capability. However, it presents some drawbacks (such as the presence of mixed sign values) and several research papers demonstrated that it is outperformed by NMF in many applications such as face recognition [25, 48]. In the following, some of the differences among these two techniques are briefly highlighted [116].

[4]The use of NMF in clustering applications will be detailed in Sect. 2.4.2.

Uniqueness. PCA is able to find the global minimum of the optimization problem, while NMF is usually trapped into local minima; this implies that the set of principal components is unique, while NMF has multiple solutions (in terms of basis and encoding matrices).

To overcome NMF nonuniqueness problem, bootstrapping techniques can be used; several executions of the factorization are performed and the most frequent solutions selected.

Ranking. Principal components are naturally ranked accordingly to the quantity of variance they explain. On the contrary, factors in NMF have no ordering and are all equally important. This causes a problem to appropriate choose the value of the rank parameter k. When PCA is applied, no specification of the value k is provided; all the eigenpairs are computed and then the most important components are selected according to the proportion of variance that one wants to preserve. Instead, when NMF is applied, the parameter k has to be specified (by user) as input parameter for the factorization. The choice of the rank value is problem dependent; usually, different factorizations are performed with different rank values and then the results are evaluated accordingly to the target of the analysis.

Orthogonality. Principal components are orthogonal directions which capture the variance in data. On the other hand, factors obtained by NMF are positive vectors that better approximate data, but they are not necessarily orthogonal. They are the bases of the hypercone containing all data and are able to preserve local data structure in this subspace. Figure 2.2 shows the principal components and the factors returned by PCA and NMF (left and right panels, respectively) when applied to nonnegative two-dimensional data matrix.

The orthogonality constraint is a desiderable property; however, this implies the presence of some negative values in the elements of principal components that, as previously highlighted, does not make sense in some contexts. The nonnegativity constraint is always violated by PCA, even when it is applied to nonnegative data. Hence, the interpretability of data is lost when moving from original data space to the reduced low-dimensionality subspace. From Fig. 2.2 (left panel) it can be observed

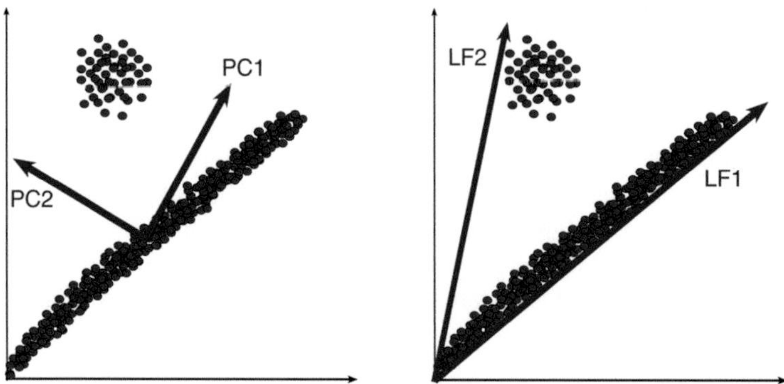

Fig. 2.2 Comparison between principal factors (*left panel*) and NMF latent factors (*right panel*)

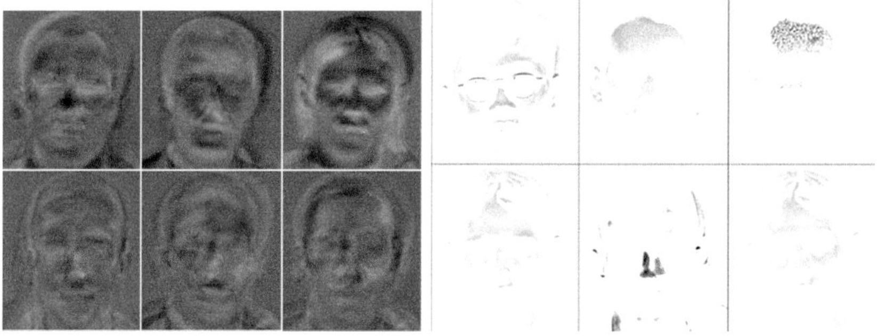

Fig. 2.3 Comparison between bases extracted with PCA (*left panel*) and NMF (*right panel*)

that, starting from samples in the positive orthant, after transforming them by PCA, samples belonging to the line assume negative values. On the contrary (right panel), NMF preserves the nonnegativity of data that leads to a part-based representation.

The interpretability of the factors is one of the strength point in NMF. The parts-based representation obtained by NMF is more intuitive and human-understandable than the holistic results of PCA. A clear example is illustrated in Fig. 2.3 in the context of facial image recognition problem [75]. PCA provides for the eigenfaces that are prototypical faces containing all kinds of facial traits (left panel), while NMF basis vectors represent particular facial traits; different kinds of eyes, noses, and mouths (right panel).

It is worth mentioning that nonnegative variants of PCA and ICA have been developed in literature [90, 95–97, 114] to overcome the difficulties of interpreting nonnegative data derived when standard PCA or ICA are used. However, these constrained variants are based on the same statistical hypotheses on the initial data as their unconstrained versions, and therefore they are applied in more specific contexts than NMF [79, 88, 89].

2.4 Constrained NMF

The key feature of NMF is to decompose the original data as combinations of parts. However, without any constraint the resulting parts could not be as intuitive as to help the analyst in a clear understanding of data. In order to be easy to understand, parts should be composed by a small number of features; however, this structural requirement must be imposed in the factorization process. This can be achieved through different possible variants of NMF which have been proposed in literature.

More specifically, the objective function

$$f(W, H) = \|X - WH\|_F^2, \tag{2.8}$$

that is minimized by the NMF factorization process[5] can be modified in several ways in order to introduce additional properties on the resulting matrices. For example, a penalty term could be added to $f(W, H)$ in order to enforce sparseness [57] as well as to enhance smoothness [35] or to improve clustering ability of NMF [31, 73, 74]. Hence, a more general objective function can be formulated

$$f(W, H) = \|X - WH\|_F^2 + \alpha J_1(W) + \beta J_2(H), \tag{2.9}$$

where the penalty terms $J_1(W)$ and $J_2(H)$ add constraints to the original problem, while the regularization parameters α and β balance the trade-off between the approximation error and additional constraints.

Penalization terms are used in order to constrain the factorization process to yield more interpretable results, so as to be more suitable for IDA. In the following, constrained variants of NMF have been reviewed.

2.4.1 Sparse NMF

Sparseness is a quality that *"refers to a representational scheme where only a few units (out of a large population) are effectively used to represent typical data vectors"* [57]. Sparse representation of hidden factors makes them easier to be interpreted because the resulting parts are structurally simple.

In fact, NMF naturally promotes a sparse representation of data. The matrices W and H describe the relationships among the original features and the latent factors, and among the latent factors and the samples, respectively. Thus, there will be many zero-entries in these matrices where such relationships are not present in data.

When the basis matrix W is sparse, basis vectors representing data subspace are sparse; thus only few features are used to describe the latent factors. This enables a part-based representation where each part is very simple and, therefore, easy to understand by the analyst. Similarly, when the encoding matrix H is sparse, then each sample is described by few (or just one) latent factors. This means that it is possible to easily explain data samples as a composition of few parts.

Sparseness is desirable because it enhances interpretability; however, it could negatively affect the accuracy of the approximation. Thus sparseness should be regulated, but this is not possible in standard NMF unless some additional constraints are added. In [57, 58], the classical NMF optimization algorithm has been modified

[5]In this chapter, we mainly consider NMF based on the error function described in (2.8), but other divergence measures could be used (e.g., generalized Kullback–Leibler divergence, α-divergence). Anyway, technical details apart, the general ideas described in the section still hold.

to include the sparseness constraint. The basic idea is to introduce a measure of sparseness of a k-dimensional vector $\mathbf{x}$ as follows[6]:

$$\text{sparseness}(x) = \frac{\sqrt{k} - (\|\mathbf{x}\|_1)/(\|\mathbf{x}\|_2)}{\sqrt{k} - 1}. \tag{2.10}$$

This measure is then used to design a projected gradient descent algorithm that controls both sparseness and nonnegativity. In essence, this algorithm essentially takes a step in the direction of the negative gradient of the cost function (2.8), and subsequently projects the solution onto the constraint space, that is the cone of nonnegative matrices with a prescribed degree of sparseness ensured by imposing the degree of sparseness to s_W and s_H for the matrices W and H, respectively.

Depending on the specific application of NMF, a desired degree of sparseness for W and H can be imposed. For example, when data samples represent images, high sparseness in both the encoding and the bases matrices is convenient. This allows to generate small *pieces* (factors) of the whole images, and few of them are used to describe each image. Differently, in a medical application where each data sample represents the symptoms of a patient and latent factors are diseases, we should expect to have a sparse encoding matrix H (because we expect patients have one or few more diseases) but W could be dense (since each disease could cause a large number of symptoms).

The prominent role of the data analyst to intelligently guide the factorization process is clear from these simple examples. Based on the questions the analyst wants to ask, and depending on the problem she needs to solve, the NMF process is modified by tuning the sparseness degree of its factors. Many variants of sparse NMF have been proposed subsequently to Hoyer's paper [58]. Some examples are sparse nonnegative matrix factorization, SNMF [39, 67, 81, 92], nonsmooth nonnegative matrix factorization [91], localNMF [37, 72], nonnegative matrix underapproximation (NMU) [41].

2.4.1.1 Nonnegative Matrix Underapproximation—NMU

More recently, Gillis and Glineur proposed the nonnegative matrix underapproximation (NMU) technique, which returns sparser representations than those obtained with classical NMF [41]. NMU is a recursive modification of the NMF algorithm obtained by imposing the upper bound constraint $WH \leq X$ to the factor matrices.

[6]The function in (2.10) yields values in the interval [0,1], where 0 indicates the minimum degree of sparseness obtained when all the elements x_i have the same absolute value, while 1 indicates the maximum degree of sparseness, which is reached when only one component of the vector x is different from zero.

Formally, given a data matrix $X \in \mathbb{R}^{m \times n}$ and a rank $1 \leq k \leq \min(m, n)$, NMU solves the following optimization problem:

$$\begin{aligned} \text{minimize}: \ & \|X - WH\|_F^2 \\ \text{subject to}: \ & WH \leq X \\ & W \geq 0, H \geq 0 \end{aligned} \tag{2.11}$$

with $W \in \mathbb{R}^{m \times k}$ and $H \in \mathbb{R}^{k \times n}$.

The basic idea is to recursively identify an optimal rank-one NMF solution $(\mathbf{w}_i, \mathbf{h}_i)$, which is easy to compute, and then apply the same technique to the residual matrix $R_{i+1} = R_i - \mathbf{w}_i \mathbf{h}_i^\top$ (being $R_1 = X$).

More precisely, we suppose that[7] $(W_{:1}, H_{1:}) = (\mathbf{w}_i, \mathbf{h}_i)$ is the solution to the underapproximation problem for X after the first iteration. Since it is an optimal rank-one NMF solution, then $W_{:1} H_{1:} \approx X \, (= R_1)$ and $W_{:1} H_{1:} \leq X$. At the second iterate, the residual matrix is $R_2 = X - W_{:1}, H_{1:}$ which is nonnegative, so it can be underapproximated by $W_{:2} H_{2:} \leq R_2$ leading to $R_3 = R_2 - W_{:2}, H_{2:}$. After k iterates

$$\begin{aligned} X &\geq W_{:1} H_{1:} + W_{:2} H_{2:} + \cdots + W_{:k} H_{k:} \\ &= [W_{:1} W_{:2} \ldots W_{:k}] \, [H_{1:}; H_{2:}; \ldots; H_{k:}] \\ &= WH \end{aligned} \tag{2.12}$$

This method has been tested on several images datasets and demonstrated that it is able to derive sparser solutions than classical NMF, thus improving the part-based representation of factors. In fact, the basis describes approximately disjoint parts of the input data. Also, differently from the classical NMF algorithms, it is possible to choose the factorization rank k during the computation; this enables the data analyst to guide the number of iterations according to the quality of the results. Finally, the optimal solution is, under some assumptions, unique [43]. An extension of NMU, which further emphasizes sparseness, has been proposed to analyze hyperspectral images [44].

2.4.2 Orthogonal NMF and Clustering Capabilities

Dimensionality reduction can be exploited for endowing NMF with clustering capabilities. The theoretical relationship between NMF (with additional orthogonal constraints on its factors), k-means, and spectral-based clustering was demonstrated [31], while the mathematical equivalence between orthogonal NMF and a weighted variant of spherical k-means was proved together with some indications about the cases in which orthogonal NMF should be preferred over k-means and spherical k-means [98].

---

[7] The symbol $W_{:i}$ denotes the ith column of W. The same applies for H.

Clustering is one of the most useful tools in IDA, since it produces a summarized view of data that helps the analyst to understand data by means of compact and informative representations of large collections of samples [7]. Many different clustering methods exist in literature, like hierarchical clustering, prototype-based clustering, and density-based clustering (just to cite the most important ones). Hierarchical clustering yields a collection of nests groups of data, while in prototype-based clustering groups are represented in a compressed form through a prototype, i.e., an element belonging to the same domain of data. Finally in density-based clustering, groups are formed in regions of data space where data are more crowded. The choice of the most appropriate method is up to the data analyst.

In the case that prototype-based clustering is a convenient method for the problem at hand, NMF could be a valid tool. NMF has been widely used in clustering applications [101, 112] where the factors W and H have been interpreted in terms of cluster centroids and cluster membership, respectively.

From a geometric point of view, columns of W are the axes of the sub-dimensional space where samples are spanned. They represent latent feature extracted from data. Vector samples are clustered according to their closeness to these basis vectors.

NMF without constraints finds a convex hull containing data points. However, [31] pointed out that adding orthogonality constraint to NMF algorithms is necessary to improve their clustering capabilities. In fact, the bases obtained from orthogonal NMF tend to point to the center of the clusters. The minimization problem in (2.8) has been modified imposing orthogonality constraint to the rows of the encoding matrix H as follows:

$$\min_{W\geq 0,\ H\geq 0} \|X - WH\|_F^2, \quad \text{s.t.} \quad HH^\top = I. \tag{2.13}$$

Orthogonality constraint on the matrix H forces samples belonging to the same cluster to be closer to same bases. In the same manner, a feature clustering can be achieved by imposing the orthogonality constraint on the columns of the basis matrix W (i.e., $W^\top W = I$).

As a natural consequence, [31] proposed a new minimization problem. Simultaneously, clustering of both features and objects (i.e., co-clustering) has been archived imposing orthonormality constraints on both columns of W and rows of H.

$$\min_{W\geq 0,\ H\geq 0} \|X - WH\|_F^2, \quad \text{s.t.}\ W^\mathrm{T} W = I,\ HH^\mathrm{T} = I. \tag{2.14}$$

In this representation, the matrix W is the clustering indicator matrix, and the rows of the matrix H are the cluster centers for the features clustering problem; while the matrix H is the clustering indicator matrix, and the columns of the matrix W are the cluster centers for the objects clustering problem. However, this double orthogonality constraint is very restrictive and it leads to a rather poor matrix low-rank approximation. Different multiplicative updates for NMF preserving orthogonality were recently proposed [23, 26, 61]. To overcome the limits of the two factor orthogonal NMF, tri-factors NMF–TNMF has been proposed. Particularly, TNMF adds an extra factor to absorb the different scales of X, W, H and to allow different number of clusters for features and objects, that is

$$X \approx USV, \tag{2.15}$$

being $X \in \mathbb{R}_{+}^{n \times m}$, $U \in \mathbb{R}_{+}^{n \times k}$, $S \in \mathbb{R}_{+}^{k \times l}$, $V \in \mathbb{R}_{+}^{l \times m}$, where the number of rows in S correspond to the number of feature-clusters k, while the number of columns to the number of objects-clusters l.

The interested reader can find a deep investigation about NMF algorithms with orthogonality constraint and their application in clustering on [68, 73, 74, 87].

2.4.3 Semi-Supervised NMF

NMF is an unsupervised machine learning algorithm, in fact it allows to *automatically* extract human-significative feature from data and to reduce the dimensionality of data. As it has been shown in the previous paragraph, classical NMF algorithms, and constrained ones, are widely used in clustering applications. They group data in a unsupervised way, but without taking in account any prior information of data. However, when class labels are available, this knowledge could be injected in the factorization process, to improve the quality of clustering. Labeling dataset could be difficult, expensive, or time consuming, and often incomplete labels are available. Semi-supervised learning methods use a large amount of unlabeled data, together with labeled data, to train the process [94].

Different algorithms have been also proposed in the context of NMF to inject a priori knowledge. This can be done extending the objective function in (2.8) to include extra terms containing the available a priori knowledge (that could be class labels associated to the samples or pairwise constraints provided by the user, which indicate data to be clustered together—*must link*—and data that have not to be clustered together—*cannot link*). Research on NMF is going in the direction of considering it an interactive tool, instead of a black box. Semi-supervised NMFs allows to modify the factorization process taking in account the knowledge of the analyst. Some examples are [11, 19–22, 51–53, 64, 71, 77, 78, 83, 109, 111, 113].

With the idea of injecting a priori knowledge into NMF process, a new algorithm have been proposed to represent data subspace by user-defined basis as prescribed by IDA [16]. This novel masked nonnegative matrix factorization (MNMF) algorithm could be used either to explain data as a composition of interpretable parts (which are actually hidden in them) and to introduce knowledge in the factorization process as it is briefly described in the following paragraph.

2.4.3.1 Masked NMF

In MNMF the structure of the basis matrix W is defined by a user-provided mask matrix. The analyst specifies the parts she is interested to discover in data and the MNMF technique extracts the subset of data that are actually represented by those parts [13].

From the vector representation of data, it is possible to observe that each sample is represented by a vector $x \in \mathbb{R}^n_+$ of n features $\{f_1, \ldots, f_n\}$. A part p is defined as a sparse vector in $\mathbb{R}^n$ where at least two components are nonzero. A feature belongs to a part iff its value is nonzero. In this way the factorization process is constrained to describe data as a linear correlation of different parts, whose features are linearly correlated among them. The structure of the part (i.e., the features set to zero, thus excluded by the part), as well as the number of parts, constitutes the a priori knowledge and is user-defined.

To obtain basis factors that are able to extract parts, the columns $\mathbf{w}_k$ in W are constrained to contain only few nonzero elements. Factors possessing this type of structure enable the elicitation of local linear relationships in subsets of data and therefore it is very useful for IDA.

A binary matrix $P \in \{0, 1\}^{n\times k}$, with the same dimensions of the basis matrix W is used as mask for the NMF problem. Particularly, the mask matrix P is used to identify the parts that the analyst would like to extract from data. This is accomplished by defining P as a set of k column vectors, where each element in a column is 1 if the corresponding feature has to be selected, 0 if it has not be considered.

To incorporate the additional constraint described above, the NMF minimization problem (2.7) has been extended to automatically impose the structure of the mask P to the basis matrix W:

$$\min_{W\geq 0,\ H\geq 0} \frac{1}{2} \|X - (P \odot W)H\|_F^2 + \frac{1}{2}\lambda \left\| P \odot \tilde{W} \right\|_F^2, \tag{2.16}$$

where $\tilde{w}_{ij} = \exp\left(-w_{ij}\right)$ and $P \in \{0, 1\}^{n\times k}$ and $\lambda \geq 0$ is a regularization parameter.

The objective function in (2.16) is composed of two terms: the first one represents a weighted modification of the classical NMF problem where the mask matrix P is used to fix the structure the basis matrix W has to possess. The second term is a penalty term used to enhance the elements w_{ij} corresponding to elements $p_{ij} = 1$. For this purpose, the exponential function has been chosen; when the value of an entry w_{ij} of W is small, it is increased by the penalty term, when it is high the penalty tends to zero. The choice of the exponential function allow us to prevent that zero values correspond to features that we want to include in the parts. The regularization parameter $\lambda \geq 0$ is used to balance the influence of the two terms.

Two updating formulas for the factors W and H have been derived as a modification of the standard multiplicative update rules of Lee and Seung, taking into account the mask constraint [16].

A query-based MNMF algorithm is used to select the parts in the query that are actually represented by data samples. However, the selected parts are generally

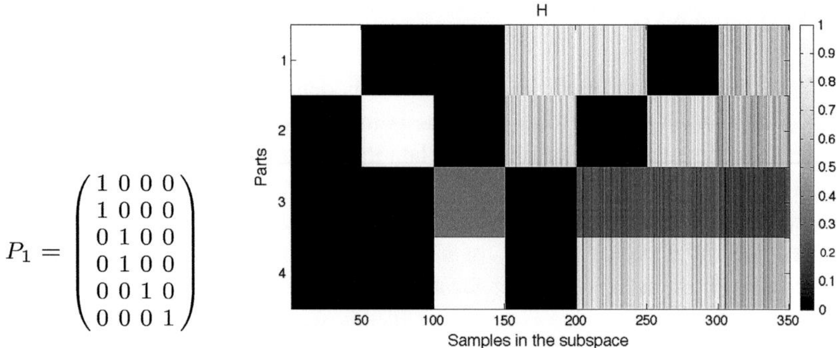

Fig. 2.4 Matrix H obtained with MNMF and P_1

represented by a subset of data only. The algorithm uses the information in the encoding matrix H to understand if samples have been correctly reconstructed by the selected parts. Particularly, the elements of each columns of the encoding matrix H codify the information needed to identify the factors (columns of W) used to reconstruct each sample in the low-rank subspace. They codify the importance of each basis vector in approximating data sample; if a coefficient is very small, then the corresponding basis vector is useless in approximating the sample; as a consequence, data sample does not contain the part represented by this basis vector.

A user-defined threshold is used to quantify the goodness of the reconstruction.

Figure 2.4 shows a simple encoding matrix H obtained with a synthetic dataset X where linear relations among feature are present. On the rows there are the parts and on the columns the samples. Colors in the image highlight the parts that have been used to reconstruct the samples. Different colors indicate how good the reconstruction have been. For instance, samples from 1 to 50 have been perfectly reconstructed using only the first part P_1. This means that, in this subset of data, a local linear relationship between the first and the second features is present. In fact, the first part, corresponding to the first column of P, selects the first and the second features.

An automatic procedure is used to extract the subset of data that are actually represented by the parts, discarding those data in the matrix X that do not find a clear representation by the parts and returning the subset of samples that contains the selected parts.

2.5 An Illustrative Example: NMF for Educational Data Mining

In order to illustrate the use of NMF for intelligently analyzing real-world data, an application of some NMF variants is discussed in the realm of educational data mining.

Educational data mining (EDM) aims at extracting useful knowledge from data coming from e-learning scenarios [100]. The different methods in EDM are designed to collect, store, and analyze data coming from learning and evaluation processes of students, in order to detect conceptual categories that are not directly observable, such as attitudes, interests, values, personality, cognitive abilities, etc.

Roughly speaking, two fundamental theories influence the choice of the appropriate method for analyzing e-learning data: classical test theory (CTT) [106] and item response theory (IRT) [82]. Both are based on the assumption that the answers to specific tests can be considered as manifestations of some skills, which are not immediately observable but can be indirectly derived from the answers.

In this context, NMF can be used to analyze data coming from student questionnaires, in order to identify the latent factors involved in the learning process. Particularly, given a data matrix X (*score matrix*) representing the answers of the students to the questions (*items*) in a test, NMF decomposes it in two factor matrices W and H, such that W encodes the relations between the latent factors (*skills*) and the questions, while H describes the abilities of the students with respect to these latent factors. The latent factors are represented with nonnegative vectors of skill, such that the skills of each student can be defined as a linear combination of these vectors.

The extracted information provides the building blocks of a learning cognitive model, which corresponds to a particular matrix, the so-called question matrix (*Q-matrix*) describing the student necessary skills to adequately answer the questionnaires [105]. Since the skills do not occur explicitly (the actual presence of a particular skill can be only hypothesized on subjective bases), the construction of a Q-matrix is a nontrivial process. To overcome this difficulties, the NMF could be used to automatically extract a Q-matrix (W) [17, 28].

In order to show the application of NMF for extracting a Q-matrix, the *SAT* dataset has been used.[8] The dataset reports the answers of 296 students to 40 questions (items) on four subjects: Mathematics (items 1–10), Biology (item 11–20), History (items 21–30), and French (items 31–40) [27] from published study guides for the SAT Subject Tests.[9]

The left panel of Fig. 2.5 represents the Q-matrix W obtained applying the standard NMF to SAT. Each row represents a skill, while the columns identify the items. The gray shade of each cell indicates the weight of each skill in characterizing the corresponding item; the lightest shades indicate the heaviest weights. As it can be observed, the first and the second skills are mainly composed by contiguous groups of items, while the last two skills are described by scattered items in the dataset.

Since the items correspond to contiguous questions on the same subject (in blocks of ten), it is possible to conclude that the skills number one and two are semantically aligned with the subjects related to the two groups of contiguous items (specifically, Mathematics, and French), while the third and fourth skills are defined by a composition of the remaining subjects (Biology and History). The latter could represent

---

[8]The dataset is available at http://alumni.cs.ucr.edu/~titus/ (accessed: March 25th 2015).

[9]http://sat.collegeboard.org.

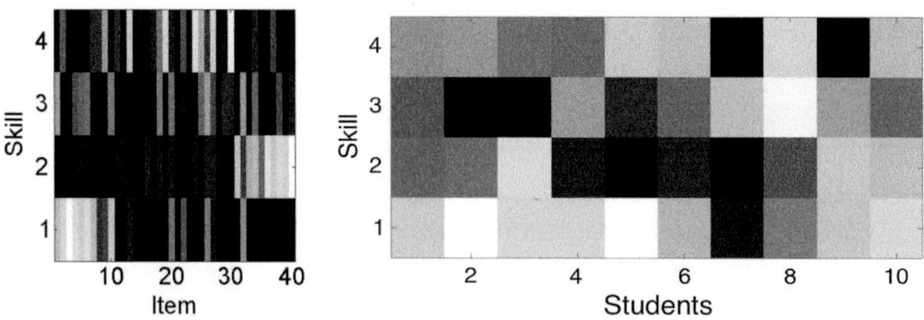

Fig. 2.5 W and H matrices obtained with the standard NMF on the SAT score matrix

"mixed abilities" that cannot be semantically framed in one of the a prior known subjects. These results could suggest a reorganization of the questions to adapt the arguments to the skills, or on the contrary they could suggest more interdisciplinary skills.

The right panel of the Fig. 2.5 illustrates the first ten columns of the matrix H resulting after factorization. This matrix identifies the degrees to which a student has acquired a particular skill. Information in H allows to highlight the skills in which each student is more or less experienced and to group the students according to their abilities, for example, to organize remedial courses.

Constrained versions of NMF could be used to add extra knowledge to the Q-matrix. For instance, by adding the sparsity constraint to NMF, the resulting Q-Matrix is more sparse, i.e., each skill is described by few items. Thus, the sparsity constraint allows to restrict the influence the items have on the skills. The most semantically relevant items can be automatically extracted, thus facilitating the analysis when the number of items is very high. On the other hand, by adding the orthogonal constraint, NMF produces a Q-matrix where skills are well separated, i.e., few items simultaneously belong to different skills. Finally, if the analyst possesses a priori information on the skills required to solve each question, she could impose these relationships by using the masked version of NMF (MNMF).

On the overall, this example gives an idea of the possibilities of NMF as an intelligent analysis tool. The different constraints can be used to reflect the needs of an analyst in the peculiar educational context.

2.6 Conclusions

In this chapter several variants of NMF algorithms have been analyzed in order to highlight their usefulness in the context of Intelligent Data Analysis.

Particularly, it has been pointed out how sparseness and orthogonality constraints have been used to modify classical NMF algorithms in order to overcome their limitations as holistic bases not corresponding with the part-based expected, and not unique decomposition. Moreover, it has been shown how to incorporate a priori knowledge in the factorization process. Semi-supervised NMF algorithm which inject meta-information about samples by adding extra matrices are also briefly introduced. A novel algorithm used to impose user-defined structure to the reduced space in which data are approximated has been also presented in some details.

The potentialities of NMF as a promising tool for IDA have been stressed: dimensionality reduction, clustering, and part-based representation. However, this chapter could be considered as a starting point in the research panorama in which NMF is no more a black box, but an interactive tool that can be driven by data analyst capabilities.

References

1. R.E. Bellman, *Adaptive Control Processes—A Guided Tour* (Princeton University Press, Princeton, 1961)
2. E. Benetos, M. Kotti, C. Kotropoulos, Applying supervised classifiers based on non-negative matrix factorization to musical instrument classification, in ICME (IEEE, 2006), pp. 2105–2108
3. E. Benetos, M. Kotti, C. Kotropoulos, Musical instrument classification using non-negative matrix factorization algorithms and subset feature selection, in Proceedings of International Conference on Acoustics, Speech and Signal Processing (ICASSP'06), vol. V (2006), pp. 221–224
4. E. Benetos et al., Comparison of subspace analysis-based and statistical model-based algorithms for musical instrument classification, in 2nd Workshop on Immersive Communication and Broadcast Systems (ICOB'05), (Berlin, Germany, 2005)
5. M. Berry et al., Algorithms and applications for approximate nonnegative matrix factorization. Comput. Stat. Data Anal. **52**(1), 155–173 (2007)
6. M. Berthold, D.J. Hand (eds.), *Intelligent Data Analysis: An Introduction*, 1st edn. (Springer, New York, 1999)
7. M.R. Berthold et al., *Guide to Intelligent Data Analysis: How to Intelligently Make Sense of Real Data*, 1st edn. (Springer, Incorporated, London, 2010)
8. R. Bierig et al., Conquering data: the state of play in intelligent data analytics (2015)
9. J.P. Brunet et al., Metagenes and molecular pattern discovery using matrix factorization. Proc. Natl Acad. Sci. 101(12), 4164–4169 (2004). doi:10.1073/pnas.0308531101. ISSN: 1091-6490
10. J.E. Burger, P.L.M. Geladi, Hyperspectral image data conditioning and regression analysis, *Techniques and Applications of Hyperspectral Image Analysis* (Wiley, Chichester, 2007)
11. D. Cai et al., Locality preserving nonnegative matrix factorization, in Proceedings of 2009 International Joint Conference on Artificial Intelligence (IJCAI'09) (2009)
12. P. Carmona-Saez et al., Biclustering of gene expression data by non-smooth non-negative matrix factorization. BMC Bioinform. **7**, 78 (2006)
13. G. Casalino, Non-negative factorization methods for extracting semantically relevant features in Intelligent Data Analysis. Ph.D. thesis, Dipartimento di Informatica, Università degli Studi di Bari (2015)
14. G. Casalino, N. Del Buono, C. Mencar, Subtractive clustering for seeding non-negative matrix factorizations. Inf. Sci. **257**, 369–387 (2014). doi:10.1016/j.ins.2013.05.038. ISSN: 0020-0255
15. G. Casalino, N. Del Buono, M. Minervini, Nonnegative matrix factorizations performing object detection and localization. Appl. Comput. Intell. Soft Comput. 2012, 15:15–15:15 (2012). doi:10.1155/2012/781987. ISSN: 1687-9724

16. G. Casalino, N. Del Buono, C. Mencar, Part-based data analysis with masked non-negative matrix factorization, in Computational Science and Its Applications–ICCSA 2014–14th International Conference, Guimarães, Portugal, 30 June-3 July 2014, Proceedings, Part VI, ed. by B. Murgante, S. Misra, A. Maria, A.C. Rocha, C. Maria Torre, J. Gustavo Rocha, M. Irene Falcão, D. Taniar, B.O. Apduhan, O. Gervasi. Lecture Notes in Computer Science, vol. 8584 (Springer, 2014), pp. 440–454. doi:10.1007/978-3-319-09153-2_33
17. G. Casalino et al., Fattorizzazioni matriciali non negative per l'analisi dei dati nell'educational data mining, in DIDAMATICA2012 (2012)
18. J. Chen, S. Feng, J. Liu, Topic sense induction from social tags based on non-negative matrix factorization. Inf. Sci. **280**, 16–25 (2014). doi:10.1016/j.ins.2014.04.048. ISSN: 0020-0255
19. Y. Chen et al., Non-negative matrix factorization for semi-supervised data clustering. Knowl. Inf. Syst. **17**(3), 355–379 (2008). doi:10.1007/s10115-008-0134-6. ISSN: 0219-1377
20. Y. Chen et al., Non-negative matrix factorization for semisupervised heterogeneous data coclustering. IEEE Trans. Knowl. Data Eng. 22(10), 1459–1474 (2010). doi:10.1109/TKDE.2009.169. ISSN: 1041-4347
21. Y. Chen et al., Incorporating user provided constraints into document clustering, in Seventh IEEE International Conference on Data Mining, ICDM 2007 (2007), pp. 103–112. doi:10.1109/ICDM.2007.67
22. Y. Cho, L.K. Saul, Nonnegative matrix factorization for semi-supervised dimensionality reduction, in CoRR (2011). abs/1112.3714
23. S. Choi, Algorithms for orthogonal nonnegative matrix factorization, in IEEE International Joint Conference on Neural Networks, 2008. IJCNN. IEEE World Congress on Computational Intelligence (IEEE, 2008), pp. 1828–1832
24. M. Chu et al., Optimality, computation, and interpretation of nonnegative matrix factorizations. SIAM J. Matrix Anal. 4–8030 (2004)
25. A. Cichocki et al., *Nonnegative Matrix and Tensor Factorizations: Applications to Exploratory Multi-way Data Analysis and Blind Source Separation* (Wiley, Chichester, 2009). ISBN 0470746661, 9780470746660
26. N. Del Buono, A penalty function for computing orthogonal non-negative matrix factorizations, in ISDA (IEEE Computer Society, 2009), pp. 1001–1005. ISBN: 978-0-7695-3872-3
27. M. Desmarais, Conditions for effectively deriving a q-matrix from data with non-negative matrix factorization. best paper award, in EDM, ed. by M. Pechenizkiy, T. Calders, C. Conati, S. Ventura, C. Romero, J.C. Stamper (2011), pp. 41–50. ISBN: 978-90-386-2537-9
28. M.C. Desmarais, B. Beheshti, R. Naceur, Item to skills mapping: deriving a conjunctive Q-matrix from data, in Intelligent Tutoring Systems (2012), pp. 454–463
29. K. Devarajan, Nonnegative matrix factorization: an analytical and interpretive tool in computational biology. PLoS Comput. Biol. **4**(7), e1000029 (2008)
30. I.S. Dhillon, S. Sra, Generalized nonnegative matrix approximations with Bregman divergences, in Proceeding of Neural Information Processing Systems (Curran Associates Inc., 2005), pp. 283–290
31. C. Ding, X. He, H.D. Simon, On the equivalence of nonnegative matrix factorization and k-means–spectral clustering, in Proceedings of the SIAM Data Mining Conference (SIAM, 2005), pp. 606–610
32. C. Ding et al., Orthogonal nonnegative matrix tri-factorizations for clustering, in Proceedings of the 12th ACM SIGKDD International Conference on Knowledge Discovery and Data Mining (ACM, 2006), pp. 126–135
33. D. Donoho, V. Stodden, When does non-negative matrix factorization give a correct decomposition into parts? in Advances in Neural Information Processing Systems, vol. 16, ed. by S. Thrun, L. Saul, B. Schölkopf (MIT Press, Cambridge, 2004)
34. K. Drakakis et al., Analysis of financial data using non-negative matrix factorization. Int. Math. Forum **3**(38), 1853–1870 (2008)
35. S. Essid, C. Févotte, Smooth nonnegative matrix factorization for unsupervised audiovisual document structuring. IEEE Trans. Multimed. **15**(2), 415–425 (2013). doi:10.1109/TMM.2012.2228474

36. U.M. Fayyad et al. (eds.), Advances in Knowledge Discovery and Data Mining (American Association for Artificial Intelligence, 1996). Chap. From data mining to knowledge discovery: an overview, pp. 1–34. ISBN: 0-262-56097-6
37. T. Feng et al., Local non-negative matrix factorization as a visual representation, in Proceedings of the 2nd International Conference on Development and Learning, ICDL'02 (IEEE Computer Society, 2002), p. 178
38. R.A. Fisher, The use of multiple measurements in taxonomic problems. Ann. Eugen. **7**(7), 179–188 (1936)
39. Y. Gao, G. Church, Improving molecular cancer class discovery through sparse non-negative matrix factorization. Bioinformatics **21**(21), 3970–3975 (2005)
40. N. Gillis, The why and how of nonnegative matrix factorization, in Regularization, Optimization, Kernels, and Support Vector Machines, ed. by M. Signoretto, J.A.K. Suykens, A. Argyriou. Machine Learning and Pattern Recognition Series (Chapman and Hall/CRC, Boca Raton, 2014)
41. N. Gillis, F. Glineur, Using underapproximations for sparse nonnegative matrix factorization. Pattern Recognit **43**(4), 1676–1687 (2010). doi:10.1016/j.patcog.2009.11.013. ISSN: 0031-3203
42. N. Gillis, D. Kuang, H. Park, Hierarchical clustering of hyperspectral images using rank-two nonnegative matrix factorization, in CoRR (2013). abs/1310.7441
43. N. Gillis, R.J. Plemmons, Dimensionality reduction, classification, and spectral mixture analysis using non-negative underapproximation. Opt. Eng. **50**(2), 027001 (2011). doi:10.1117/1.3533025
44. N. Gillis, R.J. Plemmons, Sparse nonnegative matrix underapproximation and its application to hyperspectral image analysis. Linear Algebra Appl. **438**(10), 3991–4007 (2013). doi:10.1016/j.laa.2012.04.033. ISSN: 0024-3795
45. G.H. Golub, C.F. Van Loan, *Matrix Computations*, 3rd edn. (The Johns Hopkins University Press, Baltimore, 2001)
46. G.H. Golub, A. Hoffman, G.W. Stewart, A generalization of the Eckart-Young-Mirsky matrix approximation theorem. Linear Algebra Appl. **88——89**(0), 317–327 (1987). doi:10.1016/0024-3795(87)90114-5. ISSN: 0024-3795
47. Q. Gu, J. Zhou, C.H.Q. Ding, Collaborative filtering: weighted nonnegative matrix factorization incorporating user and item graphs, in SDM (SIAM, 2010), pp. 199–210
48. D. Guillamet, J. Vitriá, Evaluation of distance metrics for recognition based on non negative matrix factorization. Pattern Recognit. Lett. **24**(9–10), 1599–1605 (2003). doi:10.1016/S0167-8655(02)00399-9. ISSN: 0167-8655
49. D. Guillamet, J. Vitriá, Non-negative matrix factorization for face recognition, in CCIA'02: Proceedings of the 5th Catalonian Conference on AI (Springer, New York, 2002), pp. 336–344
50. D.J. Hand, Intelligent data analysis: issues and opportunities, in IDA, ed. by X. Liu, P.R. Cohen, M.R. Berthold. Lecture Notes in Computer Science, vol. 1280 (Springer, New York, 1997), pp. 1–14
51. Y. He, H. Lu, S. Xie, Semi-supervised non-negative matrix factorization for image clustering with graph Laplacian. Multimed. Tools Appl. **72**(2), 1441–1463 (2014). doi:10.1007/s11042-013-1465-1. ISSN: 1380-7501
52. Y. He et al., Non-negative matrix factorization with pairwise constraints and graph Laplacian. Neural Process. Lett. pp. 1–19 (2014). doi:10.1007/s11063-014-9350-0. ISSN: 1370-4621
53. M. Heiler, C. Schnörr, Learning sparse representations by non-negative matrix factorization and sequential cone programming. J. Mach. Learn. Res. **7**, 1385–1407 (2006). ISSN: 1532-4435
54. J.H. Holmes, N. Peek, Intelligent data analysis in biomedicine. J. Biomed. Inform. **40**(6), 605–608 (2007)
55. P.K. Hopke, *Receptor Modeling in Environmental Chemistry* (Wiley, New York, 1985)
56. H. Hotelling, Analysis of a complex of statistical variables into principal components. J. Educ. Psychol. **24**, 417–441 (1933)

57. P.O. Hoyer, Non-negative matrix factorization with sparseness constraints. J. Mach. Learn. Res. **5**, 1457–1469 (2004). ISSN: 1532-4435
58. P.O. Hoyer, Non-negative sparse coding, in Neural Networks for Signal processing XII (Proceedings of IEEE Workshop on Neural Networks for Signal Processing) (2002), pp. 557–565
59. K. Huang, N.D. Sidiropoulos, A. Swami, Non-negative matrix factorization revisited: uniqueness and algorithm for symmetric decomposition. IEEE Trans. Signal Process. (TSP) **62**(1), 211–224 (2014)
60. A. Hyvärinen, Survey on Independent component analysis. Neural Comput. Surv. **2**, 94–128 (1999)
61. J. Yoo, S. Choi, Orthogonal nonnegative matrix tri-factorization for co-clustering: multiplicative updates on Stiefel manifolds. Inf. Process. Manag. **46**, 559–570 (2010)
62. J.E. Jackson, A User's Guide to Principal Components. Wiley Series in Probability and Statistics (Wiley-Interscience, Hoboken, 2003). ISBN: 0471471348
63. S. Jia, Y. Qian. A complexity constrained nonnegative matrix factorization for hyperspectral unmixing, in ICA, ed. by M.E. Davies et al. Lecture Notes in Computer Science, vol. 4666 (Springer, New York, 2007), pp. 268–276. ISBN: 978-3-540-74493-1
64. L. Jing et al., Semi-supervised clustering via constrained symmetric non-negative matrix factorization, in Brain Informatics, ed. by F. Zanzotto, et al. Lecture Notes in Computer Science, vol. 7670 (Springer, Berlin, 2012), pp. 309–319. ISBN: 978-3-642-35138-9. doi:10.1007/978-3-642-35139-6_29
65. I.T. Jolliffe, *Principal Component Analysis* (Springer, New York, 1986)
66. E. Kim, P.K. Hopke, E.S. Edgerton, Source identification of Atlanta aerosol by positive matrix facorization. J. Air Waste Manag. Assoc. 733–739 (2003)
67. H. Kim, H. Park, Sparse Non-negative matrix factorizations via alternating non-negativity-constrained least squares for microarray data analysis. Bioinformatics **23**(12), 1495–1502 (2007). doi:10.1093/bioinformatics/btm134. ISSN: 1367-4803
68. C. Lazar, A. Doncescu, Non negative matrix factorization clustering capabilities; application on multivariate image segmentation, in CISIS, ed. by L. Barolli, F. Xhafa, H.-H. Hsu (IEEE Computer Society, 2010), pp. 924–929. ISBN: 978-0-7695- 3575-3
69. D.D. Lee, H.S. Seung, Algorithms for non-negative matrix factorization, in *Advances in Neural Information Processing Systems*, vol. 13, ed. by T.K. Leen, T.G. Dietterich, V. Tresp (MIT Press, Cambridge, 2001), pp. 556–562
70. D.D. Lee, H.S. Seung, Learning the parts of objects by non-negative matrix factorization. Nature **401**(6755), 788–791 (1999). doi:10.1038/44565. ISSN: 0028-0836
71. H. Lee, J. Yoo, S. Choi, Semi-supervised nonnegative matrix factorization. IEEE Signal Process. Lett. **17**(1), 4–7 (2010). doi:10.1109/LSP.2009.2027163. ISSN: 1070-9908
72. S.Z. Li et al., Learning spatially localized, parts-based representation. Comput. Vis. Pattern Recognit. **1**, 207–212 (2001). doi:10.1109/CVPR.2001.990477
73. T. Li, C. Ding, The relationships among various nonnegative matrix factorization methods for clustering, in Proceedings of the Sixth International Conference on Data Mining, ICDM'06 (IEEE Computer Society, Washington, 2006), pp. 362–371. ISBN: 0-7695-2701-9
74. T. Li, C.H.Q. Ding, Nonnegative matrix factorizations for clustering: a survey, *Data Clustering: Algorithms and Applications* (CRC Press, Boca Raton, 2013)
75. Z. Lihong, G. Zhuang, X. Xu, Facial expression recognition based on PCA and NMF, in 7th World Congress on Intelligent Control and Automation, WCICA 2008 (2008), pp. 6826–6829. doi:10.1109/WCICA.2008.4593968
76. H. Liu, H. Motoda, Computational Methods of Feature Selection. Chapman & Hall/CRC Data Mining and Knowledge Discovery Series. (Chapman & Hall/CRC, Boca Raton, 2007). ISBN: 1584888784
77. H. Liu, Z. Wu, Non-negative matrix factorization with constraints, in AAAI, ed. by M. Fox, D. Poole (AAAI Press, 2010)
78. H. Liu et al., Constrained nonnegative matrix factorization for image representation. IEEE Trans. Pattern Anal. Mach. Intell. **34**(7), 1299–1311 (2012)

79. P. Liu et al., The application of principal component analysis and nonnegative matrix factorization to analyze time-resolved optical waveguide absorption spectroscopy data. Anal. Methods **5**(17), 4454–4459 (2013)
80. W. Liu, N. Zheng, Non-negative matrix factorization based methods for object recognition. Pattern Recognit. Lett. **25**(8), 893–897 (2004). doi:10.1016/j.patrec.2004.02.002. ISSN: 0167-8655
81. W. Liu, N. Zheng, X. Lu, Non-negative matrix factorization for visual coding, in Proceedings of IEEE International Conference on Acoustics, Speech, and Signal Processing (ICASSP'03), vol. 3 (2003), pp. 293–296
82. F.M. Lord, A Theory of Test Scores (1952)
83. N. Lyubimov, M. Kotov, Non-negative matrix factorization with linear constraints for single-channel speech enhancement, in INTERSPEECH, ed. by F. Bimbot et al. (ISCA, 2013), pp. 446–450
84. W.K. Ma et al., A signal processing perspective on hyperspectral unmixing: insights from remote sensing. IEEE Signal Process. Mag. **31**(1), 67–81 (2014). doi:10.1109/MSP.2013.2279731
85. M.W. Mahoney, P. Drineas, CUR matrix decompositions for improved data analysis. Proc. Natl. Acad. Sci. **106**(3), 697–702 (2009). doi:10.1073/pnas.0803205106
86. E. Mejía-Roa et al., BioNMF: a web-based tool for nonnegative matrix factorization in biology. Nucleic Acids Res. **36**, 523–528 (2008)
87. A. Mirzal, Clustering and latent semantic indexing aspects of the nonnegative matrix factorization, arXiv preprint arXiv:1112.4020 (2011), pp. 1–28
88. A. Montanari, E. Richard. Non-negative principal component analysis: message passing algorithms and sharp asymptotics, in CoRR (2014). abs/1406.4775
89. B. Ng, R. Abugharbieh, M.J. McKeown, Functional segmentation of fMRI data using adaptive non-negative sparse PCA (ANSPCA), in Medical Image Computing and Computer-Assisted Intervention (MICCAI) (Springer, London, 2009), pp. 490–497
90. E. Oja, M. Plumbley, Blind separation of positive sources using non-negative PCA, in Proceedings of 4th International Symposium on Independent Component Analysis and Blind Signal Separation (ICA2003) (2003), pp. 11–16
91. A. Pascual-Montano et al., Nonsmooth nonnegative matrix factorization (nsNMF). IEEE Trans. Pattern Anal. Mach. Intell. **28**(3), 403–415 (2006). doi:10.1109/TPAMI.2006.60. ISSN: 0162-8828
92. V.P. Pauca, J. Piper, R.J. Plemmons, Nonnegative matrix factorization for spectral data analysis. Linear Algebra Appl. **416**(1), 29–47 (2006). doi:10.1016/j.laa.2005.06.025. ISSN: 0024-3795
93. K. Pearson, On lines and planes of closest fit to systems of points in space. Philos. Mag. **2**(6), 559–572 (1901)
94. N.N. Pise, P. Kulkarni, A survey of semi-supervised learning methods, in Computational Intelligence and Security (IEEE Computer Society, 2008), pp. 30–34. ISBN: 978-0-7695-3508-1
95. M. Plumbley, Algorithms for non-negative independent component analysis. IEEE Trans. Neural Netw. **14**(3), 534–543 (2003)
96. M. Plumbley, Conditions for non-negative independent component analysis. IEEE Signal Process. Lett. **9**(6), 177–180 (2002)
97. M.D. Plumbley, E. Oja, A nonnegative PCA algorithm for independent component analysis. IEEE Trans. Neural Netw. **15**(1), 66–76 (2004)
98. F. Pompili et al., Two algorithms for orthogonal nonnegative matrix factorization with application to clustering, in CoRR (2012). abs/1201.0901
99. B. Ribeiro et al., Extracting discriminative features using nonnegative matrix factorization in financial distress data, in ICANNGA, ed. by M. Kolehmainen, P.J. Toivanen, B. Beliczynski. Lecture Notes in Computer Science, vol. 5495 (Springer, New York, 2009), pp. 537–547. ISBN: 978-3-642-04920-0

100. C. Romero, S. Ventura, Educational data mining: a review of the state of the art. Trans. Syst. Man Cybern. Part C **40**(6), 601–618 (2010). ISSN: 1094-6977
101. F. Shahnaz et al., Document clustering using nonnegative matrix factorization. Inf. Process. Manag. **42**(2), 373–386 (2006). doi:10.1016/j.ipm.2004.11.005. ISSN: 0306-4573
102. C. Spearman, General intelligence, objectively determined and measured. Am. J. Psychol. **15**, 201–293 (1904)
103. X. Sun, Q. Zhang, Z. Wang, Face recognition based on NMF and SVM. Electron. Commer. Secur. Int. Symp. **1**, 616–619 (2009). doi:10.1109/ISECS.2009.98
104. R. Tandon, S. Sra, Sparse nonnegative matrix approximation: new formulations and algorithms. Technical report, MPI Technical report (2010)
105. K.K. Tatsuoka, Rule space: an approach for dealing with misconceptions based on item response theory. J. Educ. Meas. (1983)
106. Theory of Mental Tests. Wiley Publications in Psychology (Wiley, New York, 1950)
107. M. Turk, A. Pentland, Eigenfaces for recognition. J. Cogn. Neurosci. **3**(1), 71–86 (1991). doi:10.1162/jocn.1991.3.1.71. ISSN: 0898-929X
108. L.J.P. Van der Maaten, E.O. Postma, H.J. van den Herik, Dimensionality reduction: a comparative review (2008)
109. C. Wang et al., Non-negative semi-supervised learning, in AISTATS, JMLR Proceedings, vol. 5, ed. by D.A. Van Dyk, M. Welling, JMLR (2009), pp. 575–582
110. F. Wang et al., Community discovery using nonnegative matrix factorization. Data Min. Knowl. Discov. **22**(3), 493–521 (2011). doi:10.1007/s10618-010-0181-y. ISSN: 1384-5810
111. Y. Wang et al, Fisher non-negative matrix factorization for learning local features, in Asian Conference on Computer Vision (2004)
112. W. Xu, X. Liu, Y. Gong, Document clustering based on nonnegative matrix factorization, in Proceedings of the 26th Annual International ACM SIGIR Conference on Research and Development in Information Retrieval, SIGIR'03 (ACM, New York, 2003), pp. 267–273. ISBN: 1-58113-646-3
113. Y. Yang, B.-G. Hu, Pairwise constraints-guided non-negative matrix factorization for document clustering, in IEEE/WIC/ACM International Conference on Web Intelligence (2007), pp. 250–256. doi:10.1109/WI.2007.66
114. R. Zass, A. Shashua, Nonnegative sparse PCA, in Neural Information Processing Systems (2007)
115. Z. Zhang, Nonnegative matrix factorization: models, algorithms and applications, in DATA MINING: Foundations and Intelligent Paradigms, vol. 2, ed. by D.E. Holmes, L.C. Jain (Springer, Berlin, 2011), pp. 99–134
116. A. Zinovyev et al. Blind source separation methods for deconvolution of complex signals in cancer biology, in CoRR (2013). abs/1301.2634

Chapter 3
Automatic Extractive Multi-document Summarization Based on Archetypal Analysis

Ercan Canhasi and Igor Kononenko

Abstract The applications of matrix factorization are an important tool for text summarization. In last years, several variations of the non-negative matrix factorization (NMF) methods have found their usage in multi-document summarization (MDS). For matrix factorization to work efficiently in MDS, it is essential to show the ability of selecting the most typical data points from the given data space. In the chapter, we first describe the archetypal analysis (AA) and its weighted version and then we present the AA-based document summarization method for the two most known summarization tasks, namely the general and the query-focused MDS. Archetypal analysis, also known as the convex NMF, in contrast to other NMF methods selects distinct (archetypal) sentences and therefore leads to variability and diversity in content of the generated summaries. We conducted experiments on the data of document understanding conference. Experimental results evidence the improvement of the proposed approach over other closely related methods including ones using the NMF.

3.1 Introduction

In extractive document summarization, one of the primary goals is to get the best possible sentence selection. Common sentence selection methods based on the matrix decomposition techniques like the one based on the nonnegative matrix factorization give an averaged summary quality; however, they merely extract prototypical, characteristic, and even basic sentences. Having the method able to select distinct (archetypal) sentences and thus induces variability and diversity in produced summaries could be of a great benefit. Numerous different matrix factorization methods for document summarization are being used today and range from low-rank approx-

E. Canhasi (✉) · I. Kononenko
Faculty of Computer and Information Science,
University of Ljubljana, Tržaška cesta 25, 1000 Ljubljana, Slovenia
e-mail: ercan.canhasi@uni-prizren.com

I. Kononenko
e-mail: igor.kononenko@fri.uni-lj.si

G.R. Naik (ed.), *Non-negative Matrix Factorization Techniques*,
Signals and Communication Technology, DOI 10.1007/978-3-662-48331-2_3

imations (for example, singular value decomposition, principal component analysis, latent semantic indexing, nonnegative matrix factorization, and symmetric-NMF) to soft clustering approaches (for example fuzzy K-medoids and the EM algorithm for clustering) and hard assignment clustering methods such as K-means. Sentence selection is usually as important as the summarization itself. A good sentence selection makes the summaries finer and the document summarization system more efficient. We attempt to present a general and a query-focused extractive multi-document summarization method based on Archetypal Analysis. Such a method provides a various advantages: (i) it is an unsupervised method; (ii) it is a language independent method; (iii) it is also a graph-based method; (iv) in contrast to other factorization methods which extract prototypical, characteristic, and even basic sentences, AA and its weighted version select distinct (archetypal) sentences and thus induce variability and diversity in produced summaries; and finally (v) it readily offers soft clustering, i.e., simultaneous sentence clustering and ranking. To show the efficiency of the proposed approach, we compare it to other closely related summarization methods. We use the DUC2004 and DUC2006 datasets to test our proposed method empirically. Experimental results show that our approach significantly outperforms the baseline summarization methods and that it well compares to the most of the state-of-the-art approaches.

Contribution and Organization of the Chapter. The contribution of this chapter is threefold. First, we compile our previous findings on archetypal analysis usage in document summarization [1, 2]. Second, we extensively test the method. Third, we use the findings to provide guidelines for further developments. In Sect. 3.2, we first briefly present the related work. In Sect. 3.3, we describe the archetypal analysis method and its variations. We continue in Sect. 3.4 with the description of the proposed AA-based summarization method. Section 3.5 gives the description of the test environment and datasets we used for testing. The results are also presented in Sect. 3.5. We conclude the paper and set guidelines for further work in Sect. 3.6.

3.2 Related Work

Document summarization. In nonrestricted (general) extractive summarization, each sentence is scored with a significance value, and then sentences are ranked based on the significance score. The significance scores are commonly calculated as a combination of syntactic, semantic, and statistical characteristics. In last years, many approaches for sentence extraction building upon the algebraic methods, more precisely matrix decomposition methods have emerged. Common approaches used in MDS spread from low-rank approximations such as singular value decomposition (SVD) [3], principal component analysis (PCA) [4], latent semantic indexing (LSI/LSA)[5], non-negative matrix factorization (NMF) [6], symmetric-NMF [7], probabilistic LSA [8] to soft clustering approaches such as the EM algorithm for clustering [9] and hard assignment clustering methods such as K-means [7], even to eigen decomposition which are also known as the graph-based methods [10].

Extracting summary for answering user's stated information need expressed by a given topic or query is the main constrain of the query focused summarization task. In it, the query should be incorporated into very nature of the summarization method. Many methods for the generic summarization can be extended to incorporate the query information, including singular value decomposition (SVD) [11], latent semantic indexing (LSA) [5], nonnegative matrix factorization (NMF) [12], and symmetric-NMF [7]. The graph-based ranking algorithms were also used in query-focused summarization when it became a popular research topic. For instance, a topic-sensitive version of LexRank is proposed by [13].

Archetypal Analysis and Weighted AA. Archetypal analysis originates from the Greek word "archetype" which stands for the prototype based on which others are formed. As follows, identifying these prototypes, archetypes, within a given dataset is the primary purpose of archetypal analysis. Archetypal analysis is a general unsupervised learning technique that can be used in many different research areas, such as in economics, text mining, statistics, and in pattern recognition [14]. The problem of identifying the archetypes in a dataset can be interpreted as a problem of discovering a few instances (archetypes) in a set of multivariate dataset so that all the data points can be properly defined as convex combinations of the archetypes. The usefulness of AA in feature extraction and dimensional reduction for a large variety of machine learning problems taken from computer vision, neuroimaging, chemistry, text mining, and collaborative filtering, is vastly presented by [15]. The concept of AA was originally formulated in [16]. Cutler and Breiman (1994) formalized AA as the problem of learning the convex hull of a dataset, and solved it using alternating non-negative least squares method, see Fig. 3.1b. They defined the problem as a nonlinear least squares problem and presented an alternating minimizing algorithm to solve it. The weighted version of the Archetypal analysis was first introduced in a paper by [17]. Appealingly, archetypal analysis is closely related to popular matrix factorization approaches such as (convex) non-negative matrix factorization (NMF)[18], even though all these formulations were independently invented around the same time. Archetypal analysis indeed produces sparse representations of the data points, by approximating them with convex combinations of archetypes; it also provides a non-negative factorization when the data matrix is non-negative.

Our previous works. It is important to note that the archetypal analysis approach to document summarization has been investigated in our previous works [1, 2]. There are some minor differences between this and our previous works: (1) In [1], the general (unrestricted) document summarization problem is treated by means of a plain archetypal analysis problem. Sentences from original documents have been modeled as the joint matrix of a content and a similarity graph. The obtained matrix is then further analyzed by AA in order to simultaneously cluster and rank the original sentences. (2) In [2], the query-focused summarization problem is formalized as a weighed version of the archetypal analysis problem. Input data, i.e., sentences from the original documents and the given query, are modeled as a multi-element graph of documents, sentences, terms, and query. This paper, differently from our previous

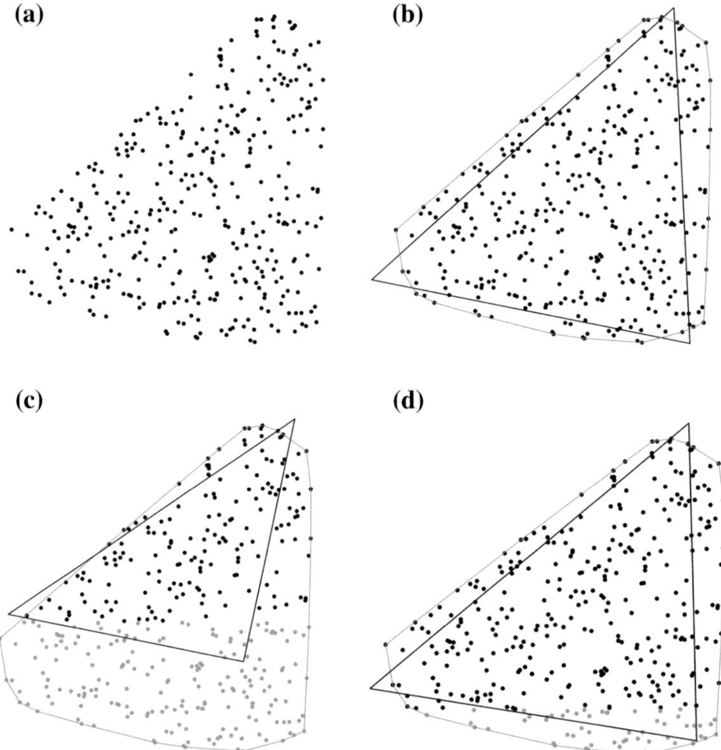

Fig. 3.1 While the archetypal analysis estimates the convex hull of a set of data (**b**), the weighted archetypal analysis approximates the weighted convex hull, with respect to points weight (**c**, **d**). Here *gray data points* weight 0.3 in (**c**) and 0.9 in (**d**) while *black data points* weight 1.0 in both. As expected, depending on the points weights the corresponding archetype changes its position inside the weighted dataset boundary. **a** Toy data set in two dimensional space. **b** Convex hull and hull approximated with 3 archetypes. **c** Three weighted archetypes; weight $= 0.3$. **d** Three weighted archetypes; weight $= 0.9$

work, does not treat only the general or query-focused summarization task but rather proposes a unified method for treating the general and query-focused summarization tasks while emphasizing the aspect of archetypal analysis.

3.3 Archetypal Analysis

In order to motivate the (w)AA algorithm and to trace the development process to the point of (w)AA, in this section we first summarize the archetypal analysis as presented in its original form [16] and then we briefly present the weighted version of AA (wAA) mainly inherited from [17].

Let as consider a matrix $X = [x_1, ...x_n]$ in $\mathbb{R}^{n\times m}$ representing a multivariate dataset with n observations and m variables. Archetypal analysis for given archetype number $p \ll n$ factorizes a given matrix X into stochastic matrices $P \in \mathbb{R}^{n\times p}$ and $Q \in \mathbb{R}^{n\times p}$ as defined by Eq. (3.1)

$$X \approx PQ^T X \tag{3.1}$$

More exactly, the archetypal problem is to calculate two matrices Q and P which minimize the residual sum of squares

$$\begin{aligned} RSS(k) &= \|X - PV^T\|^2 \text{ with } V = X^T Q \\ \text{s.t.} &\sum_{j=1}^{p} P_{ij} = 1,\, P_{ij} \geq 0; \sum_{i=1}^{m} Q_{ij} = 1,\, Q_{ij} \geq 0 \end{aligned} \tag{3.2}$$

The feature matrix V is ensured to be a convex combination of the archetypes via the constraints $\sum_{i=1}^{p} P_{ij} = 1$ and $P_{ij} \geq 0$. Similarly, the constraints $\sum_{i=1}^{m} Q_{ij} = 1$ and $Q_{ij} \geq 0$, enforce that each archetype is a meaningful combination of data points. $\|\cdot\|^2$ denotes the Euclidean matrix norm.

Equation (3.2) defines the bases of the estimation algorithm presented by pioneering AA paper [16]. It alternates between finding the best P for given archetypes V and finding the best archetypes V for given Q; where at each step many convex least squares problems are solved until the overall RSS is reduced successively.

The Algorithm 1 express the inclusive framework for archetypal analysis. The original archetypal problem as defined by Eq. (3.1), requires that each data point and hence each residual participate to the minimization with the same weight. Assuming W is a complementing $n \times n$ square weight matrix, the weighted version of the archetypal problem [17] can be formulated as the minimization of:

$$RSS(k) = \|W(X - PV^T)\|^2 \text{ with } V = X^T Q \tag{3.3}$$

Since weighting the residuals is equivalent to weighting the dataset:

$$\begin{aligned} W(X - PY^T) &= W(X - P(X^T Q)^T) = W(X - PQ^T X) = WX - W(PQ^T X) \\ &= WX - (WP)(Q^T W^{-1})(WX) = \hat{X} - \hat{P}\hat{Q}\hat{X} = \hat{X} - \hat{P}\hat{V}^T \end{aligned}$$

the problem can be rewritten as minimizing

$$RSS(k) = \|\hat{X} - \hat{P}\hat{V}\|^2 \text{ with } \hat{V} = \hat{Q}\hat{X} \text{ and } \hat{X} = WX \tag{3.4}$$

Using the original algorithm in calculating the weighted AA is possible based on this reformulation with the supplementary pre-processing step to calculate $\hat{X}$ and the additional post-processing step to recalculate P and Q for the dataset X given the archetypes $\hat{V} = Q\hat{X}$. The weight matrix W can represent different intentions, see Fig. 3.1c.

Algorithm 1 Archetypal Analysis algorithm

Input: Input matrix X and the number of archetypes p;

Output: Factorized sentences P and Q, where the archetypes themselves are the matrix product $Y = XQ$;

1: Pre-processing: scale data;

2: Initialization: initialize Q in a way the constrains are satisfied to calculate the starting archetypes V, also to ensure the Breiman's point on choosing initial mixtures (archetypes) we use the following method. The method proceeds by randomly selecting a data point as an archetype and selecting subsequent data points x_i (archetypes) the furthest away from already selected ones x_j. Such a new data point is selected according to

$$a^{new} = \max_i \left\{ \sum_j \|x_i - x_j\|, j \in Q \right\}$$

where $\|\cdot\|$ is a given norm and Q is a set of indexes of current selected points.

3: Sort the archetypes in the decreasing order of significance, i.e., order the columns of matrix A based on values of Pa_i;

4: Repeat while a stopping criterion is not met, i.e., stop when RSS is small enough or the maximum number of iteration is reached:

a: Find best P for the given set of archetypes V, i.e., solve n convex least squares problems, where $i = 1, ..., n$

$$\min_{P_i} = \frac{1}{2}\|X_i - VP_i\|^2 \text{ s.t. } \sum_{j=1}^{p} P_{ij} = 1, P_{ij} \geq 0.$$

b: Recalculate archetypes $\hat{V}$ by solving a system of linear equations $X = P\hat{V}^T$.

c: Find best Q for the given set of archetypes $\hat{V}$, i.e., solve p convex least squares problems where $j = 1, ..., p$

$$\min_{Q_j} = \frac{1}{2}\|\hat{V}_j - XQ_j\|^2 \text{ s.t. } \sum_{i=1}^{m} Q_{ij} = 1, Q_{ij} \geq 0.$$

d: Recalculate archetypes $V = X^T Q$.

e: Recalculate RSS.

5: Post-processing: rescale archetypes.

Example 1. For the sake of demonstrating the AA and NMF in use, we illustrate the following example. Assuming X is an 12×12 matrix, NMF decomposes it into two non-negative matrices, N and M, and AA factorizes it into two stochastic matrices, Q and P as shown in Fig. 3.2.

Convergence: Cutler and Breiman (1994) in their pioneering work demonstrated that alike many alternating optimization algorithms, their algorithm results in a fixed point of an appropriate transformation. However, there is no guarantee that this will be a global minimizer of RSS. For further details on convergence see [16].

Discussions and Relations. Recently, different matrix factorization methods such as PCA/SVD and NMF have been prosperously utilized in MDS. Therefore, we

(a)

$$\underbrace{\begin{bmatrix} 1 & 2 & 3 \\ 4 & 5 & 6 \\ 7 & 8 & 9 \\ 10 & 11 & 12 \end{bmatrix}}_{X_{nmf}} \approx \underbrace{\begin{bmatrix} 0 & 3.7423 \\ 5.9965 & 2.8865 \\ 11.9938 & 2.0307 \\ 17.9910 & 1.1749 \end{bmatrix}}_{N} \times \underbrace{\begin{bmatrix} 0.5384 & 0.5765 & 0.6146 \\ 0.2673 & 0.5345 & 0.8018 \end{bmatrix}}_{M} = \underbrace{\begin{bmatrix} 1.0004 & 2.0004 & 3.0004 \\ 4.0000 & 5.0000 & 6.0000 \\ 7.0000 & 8.0000 & 9.0000 \\ 10.0000 & 11.0000 & 12.0000 \end{bmatrix}}_{\hat{X}_{nmf}}$$

(b)

$$\underbrace{\begin{bmatrix} 1 & 2 & 3 \\ 4 & 5 & 6 \\ 7 & 8 & 9 \\ 10 & 11 & 12 \end{bmatrix}}_{X_{AA}} \approx \underbrace{\begin{bmatrix} 0 & 1 \\ 0.3333 & 0.6667 \\ 0.6667 & 0.3333 \\ 1 & 0 \end{bmatrix}}_{P} \times \underbrace{\begin{bmatrix} 0 & 0 & 0 & 1 \\ 1 & 0 & 0 & 0 \end{bmatrix}}_{Q^T} \times \underbrace{\begin{bmatrix} 1 & 2 & 3 \\ 4 & 5 & 6 \\ 7 & 8 & 9 \\ 10 & 11 & 12 \end{bmatrix}}_{X_{AA}} = \underbrace{\begin{bmatrix} 1 & 2 & 3 \\ 4 & 5 & 6 \\ 7 & 8 & 9 \\ 10 & 11 & 12 \end{bmatrix}}_{\hat{X}_{AA}}$$

Fig. 3.2 Factorization examples from NMF and AA. **a** NMF factorization outcomes. **b** AA factorization outcomes

believe, it is in interest of the reader to study the relation of archetypal analysis and other similar factorization methods. In terms of optimization problems, NMF and AA are merely special cases of a more general problem P_G. Given any matrix $X \in \mathbb{R}_+^{n\times m}$ and any positive integer p, the problem P_G is to find the best non-negative factorization $P \approx L_1 L_2$ (with $L_1 \in \mathbb{R}_+^{n\times p}$, $L_2 \in \mathbb{R}_+^{p\times m}$), i.e.,

$$(L_1 L_2) = \min_{L_1 L_2} \|P - L_1 L_2\|^2 \tag{3.5}$$

Here, we outline NMF and AA from the aspect of the specificity of constants involved in the problem:

3.AA :

$$(Q, P) = \min_{Q,P} \|X - PQ^T X\|^2$$

s.t. Q is stochastic
P is stochastic

2.NMF :

$$(Q, P) = \min_{Q,P} \|X - QP\|^2$$

s.t. Q is non-negative
P is non-negative

where P and Q are output or decomposition matrices, X is input or matrix to decompose, while stochastic, binary, and non-negative are constrains involved in each optimization problem.

3.4 The Proposed Approach

The framework of archetypal analysis is shown in Fig. 3.3. A raw input of the algorithms is a document set which is then decomposed into the sentence set. When aside the document set a query is provided, our method switches to the query-focused summarization, otherwise the summarization is the general one. Since the aim of the proposed approach is extractive content selection, then the sentence shall be considered the summary composition item. The archetypal analysis-based summarization method mainly includes the following steps:

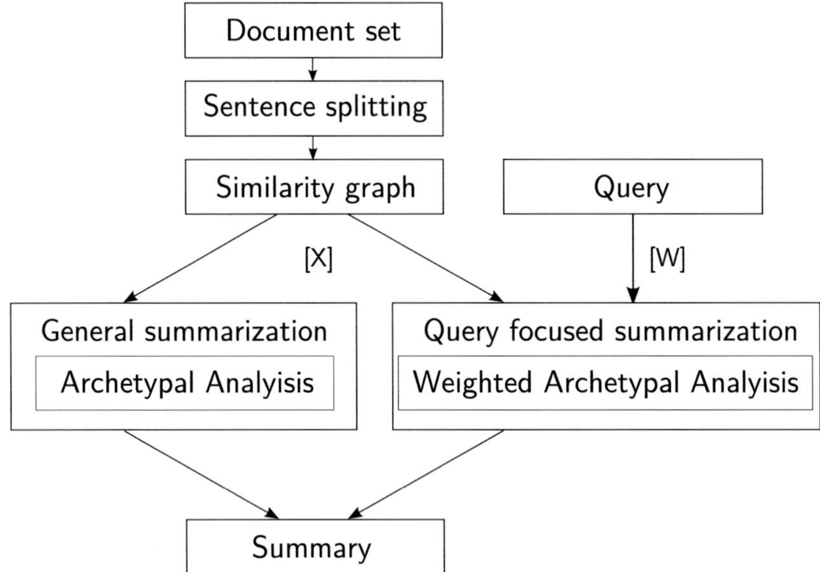

Fig. 3.3 Framework of our MDS method using AA

Step 1. Building a similarity graph and producing the weighting matrix given the query . It is a rather simple step in Fig. 3.3 before simultaneous sentence clustering and raking via AA(wAA).

Step 2. Having a soft clustering and ranking. This step aims to build a unified sentence selection model for incorporating archetypal analysis, weighted archetypal analysis, and query-relevant information. The goal of this step is to improve the coverage of summary content by the partition of subtopics through archetypes. The clustering process is dependent on the archetypal structure of a similarity matrix. Yet, another goal of this step is to provide sentence ranking.

Step 3. Ordering subtopics and selecting sentences. This process depends on the rank score from previous step and the new redundancy removal strategy.

The summary is iteratively produced by choosing the sentences from the archetypes (subtopics) with the highest rank scores based on the optimization sentence selection strategy.

Constructing the similarity graph. Given a set of documents, let an undirected and weighted similarity graph $G = (Vr, Ed)$ to reflect the relationships between sentences in document set. Vertex set $Vr = x_1, \ldots, x_n$ denotes the sentence set, each vertex x_i in Vr is a sentence. Ed is the set of edges, which is a subset of $Vr \times Vr$. Each edge between two vertices x_i and x_j carries a non-negative pairwise similarity weight $w_{ij} (i \neq j)$ in Ed. Here, we just use the standard cosine measure to compute the similarity values, which is denoted as $w_{ij} = sim(x_i, x_j)$. We remove the stop words in each sentence, and stem the remaining words. The weight associated with term t is calculated with the $tf_t * isf_t$ formula, where tf_t is the frequency of term t in the sentence and isf_t is the inverse sentence frequency of term t, i.e., $1 + log(N/n_t)$, N is the total number of sentences and n_t is the number of the sentences containing term t. Then $sim(x_i, x_j)$ is computed according to the normalized inner product of the corresponding term vectors.

Sentence soft clustering and ranking. The automatic multi-document summarization approach based on archetypal analysis presented in this chapter consist of two parts working simultaneously: (1) sentence soft clustering into (weighted) archetypes; and (2) sentence probabilistic ranking. Documents can be represented as the set of archetypes, which views the main topic from different aspects. Hence, to improve the coverage of summary content, we propose to use (w)AA which readily combines the soft clustering with the ranking to have a optimal sentence extraction.

Given the sentence similarity graph via the Algorithm 2. sentences can be soft clustered and ranked into archetypes or weighted archetypes depending on whether the query is provided or not:

Algorithm 2 (weighted) Archetypal Analysis-based sentence clustering and ranking

Input: Sentence similarity graph represented as the matrix X, weight matrix W (optional);
Output: Archetypes (subtopics) represented by matrices Q and P;
Perform the archetypal analysis or the weighted AA on matrix X given the W

1: Estimate the factorization matrices Q and P;
2: For each archetype i calculate its significance Pa_i, i.e., the sum of values in the corresponding column of the matrix $X^T Q$, $Pa_i = \sum_{j=1}^{m} X^T Q_{i,j}$;
3: Sort the archetypes in the decreasing order of significance, i.e., order the columns of matrix Q based on values of Pa_i;

Our purpose in Algorithm 2. is to cluster sentences into (weighted) archetypes and subsequently in the next step to select the sentences with the highest archetype membership weights. Here, the significant sentences are more likely to be clustered into (weighted) archetypes with high significance. Since the sentences with the higher membership values are ranked higher, the sentences selected by the Algorithm 3. are the most central ones.

Ordering subtopics and selecting sentences. Based on the archetypal analysis-based probabilistic sentence clustering and ranking, we designed a new optimization algorithm for sentence selection shown in Algorithm 3.

Algorithm 3 Ordering archetypes and selecting sentences

Input: Sorted archetypes (subtopics) $Q_1, ..., Q_p$ and sentences set $Ss = \{x_i | i = 1, 2, ...n\}$
Output: summary SUM;

1: Start with the most significant archetype and select sentences for next step in the order according to their values in Q. That is, sentences with the highest archetype membership value in each archetype (column of matrix Q) are selected and if the summary length is not met then the extraction step continues with the second highest values, and so on.;
2: Let Q_i be the top 1 archetype;
3: For Q_i, suppose x_i is the highest ranked sentence. Sentence x_i is moved from Ss to SUM, and then the redundancy penalty is imposed to the overall rank score of each sentence linked with $x_i \in Q_i$ as follows: for each sentence $xj \in Q_i$, its rank score $RScore(x_j)$ is computed by $RScore(x_j) = RScore(x_j) * (1 - S_{ji})^t$, where $t > 0$ is the exponent decay factor. The larger t is, the greater penalty is imposed to the overall rank score. If t = 0, no diversity penalty is imposed at all; In our experiments, we set $t = 3$;

The Algorithm 3 is based on the idea that extracting the summary sentences from the different subtopics helps to understand the topic from different aspects; the overall rank score of less informative sentences overlapping with the sentences in summary is iteratively decreased. Here, redundancy removal is also the key step of content selection. Finally, the sentence with the highest rank score in the most important subtopic is chosen to produce the summary until satisfying the summary length limit.

3.5 Experiments

In our experiments for summary evaluation, we used Recall-Oriented Understudy for Gisting Evaluation (ROUGE) package [19], which compares various summary results from several summarization methods with summaries generated by humans. It measures how well a machine summary overlaps with human summaries using N-gram co-occurrence statistics, where an N-gram is a contiguous sequence of N words.

The experiments are conducted on the three summarization tasks to evaluate our summarization framework. The DUC04 dataset is used for the general (unrestricted) summarization task. As for the query-focused summarization, the DUC05 and the DUC06 datasets are used. The compact view of the datasets can be found in Table 3.1.

General summarization. For the general summarization, we use DUC04 as the experimental data. We observe through experiments that the summary results generated by our method are highest for the archetype number $p = 16$. We work with the following widely used or recently published methods for general summarization as the baseline systems to compare with our proposed method AASum: (1) BaseLine: the baseline method used in DUC04; (2) Best-Human: the best human-summarizers performance (3) System-65: The best system-summarizer from DUC04; (4) System-

Table 3.1 Experimental data description

	DUC04	DUC05	DUC06
Type of summarization	General	Query	Query
Cluster #	50	50	50
Documents # per set	10	25–50	25
Summary length	665 bytes	250 words	250 words

Table 3.2 General summarization task

	ROUGE-1	ROUGE-2	ROUGE-SU
Baseline	0.3242(9)	0.0641(9)	–
Best-Human	0.4182(1)	0.1050(1)	–
System-65	0.3822(2)	0.0921(2)	0.1332(1)
System-35	0.3708(3)	0.0834(5)	0.1273(4)
LexRank	0.3784(4)	0.0857(4)	0.1312(2)
Centroid	0.3672(7)	0.0737(6)	0.1251(5)
LSA	0.3414(8)	0.0653(8)	0.1194(7)
NMF	0.3674(6)	0.0726(7)	0.1291(3)
AASum	0.3706(5)	0.0871(3)	0.1229(5)

Evaluation of the methods on the DUC04 dataset. Remark: "–" indicates that the method does not officially report the results

35: The second best system-summarizer from DUC04; (5) Lex-PageRank: by calculating the eigenvector centrality given the sentence to sentence similarity graph the method extracts the most significant sentences; (6) Centroid: which selects sentences based on the positional value and the first sentence overlap and the distance to centroid value; (7) Latent Semantic Analysis (LSA): by employing latent semantic analysis method extracts "semantically" important sentences; (8) Non-negative Matrix Factorization (NMF): the method calculates NMF on the sentence-term matrix and based on factorization matrices extracts the high ranked sentences; and (9) AASum: Archetypal analysis summarization system of the sentence similarity graph.

Table 3.2 shows the ROUGE scores of different methods using DUC04. The higher ROUGE score indicates the better summarization performance. The number in parentheses in each table slot shows the ranking of each method on a specific dataset. As indicated in [6], LSA and NMF are two competing matrix decomposition techniques for the task of MDS. From Table 3.2, we can see that NMF shows better performance than LSA. This is in consistency with results reported in [6], and it can be mainly contributed to the property of NMF to select more meaningful sentences by using the more intuitively interpretable semantic features. Our proposed approach shows even better performance. This is because it uses the archetypal analysis to detect the archetypal structure which can cluster and rank sentences more effectively than above-mentioned approaches.

Table 3.3 Query-focused summarization task

	DUC2005			DUC2006		
Summarizers	ROUGE-1	ROUGE-2	ROUGE-SU	ROUGE-1	ROUGE-2	ROUGE-SU
Avg-Human	0.4417(1)	0.1023(1)	0.1622(1)	0.4576(1)	0.1149(1)	0.1706(1)
Avg-DUC05/06	0.3434(7)	0.0602(7)	0.1148(7)	0.3795(7)	0.0754(7)	0.1321(7)
System-15/24	0.3751(4)	0.0725(4)	0.1316(4)	0.4102(2)	0.0951(2)	0.1546(2)
System-4/12	0.3748(5)	0.0685(5)	0.1277(5)	0.4049(4)	0.0899(3)	0.1476(4)
SNMF	0.3501(6)	0.0604(6)	0.1229(6)	0.3955(5)	0.0854(6)	0.1398(5)
NMF	0.3110(8)	0.0493(8)	0.1009(9)	0.3237(9)	0.0549(8)	0.1061(8)
LSA	0.3046(9)	0.0407(9)	0.1032(8)	0.3307(8)	0.0502(9)	0.1022(9)
Biased-Lex	0.3861(2)	0.0753(2)	0.1363(3)	0.3899(6)	0.0856(5)	0.1394(6)
wAASum	0.3790(3)	0.0735(3)	0.1365(2)	0.4075(3)	0.0872(4)	0.1531(3)

Evaluation of the methods on the DUC05 and DUC06 dataset

Query-focused summarization. We conduct our query-focused summarization experiments on DUC05 and DUC06 datasets since the main task of both was query-focused summarization. We compare our system with some effective, widely used and recently published systems: (1) Avg-Human: average human-summarizer performance on DUC05/06; (2) Avg-DUC05/06: average system-summarizer performance on DUC05/06; (3) System-15/24: The best system-summarizer from DUC05/06; (4) System-4/12: The second best system-summarizer from DUC05/06; (5) SNMF: the method forms the sentence to sentence similarity matrix then clusters the sentences via the symmetric non-negative matrix factorization, and finally extracts the sentences based on the clustering result. (6) Non-negative Matrix Factorization (NMF); (7) Latent Semantic Analysis (LSA); (8) Biased-LexRank: the method first constructs a sentence connectivity graph based on the cosine similarity and then selects important sentences biased toward the given query based on the concept of eigenvector centrality; and (9) wAASum: weighted Archetypal analysis summarization system of the sentence similarity graph;

The empirical results are reported in Table 3.3. The results show that on DUC05 and DUC06, our method ranks among the best top three systems and apparently outperforms the other matrix decomposition systems.

3.6 Conclusion and Further Work

In this paper, we presented an extractive summarization framework based on two alternative models able to integrate the archetypal analysis-based sentence selection, namely the plain archetypal analysis sentence clustering and ranking for general and the weighted archetypal analysis sentence selection for the query-focused document summarization.

The main contributions of the chapter are:(1) A general and query-focused document summarization method which selects important sentences from a given document set while reducing redundant information is presented, (2) The problem of general and query document summarization is formalized as the weighted Archetypal Analysis problem that takes into account relevance, information coverage and diversity, (3) This chapter has found that (w)AA-based summarization approach is an effective summarization method. Experimental results on the DUC2004, DUC2005, and DUC2006 datasets demonstrate the effectiveness of the proposed approach, which compares well to most of the existing matrix decomposition methods, and (4) The chapter proposes a unified method for treating the general and query-focused summarization tasks while emphasizing the aspect of archetypal analysis.

We believe the performance of the presented method would possibly be further improved. There are many potential directions for improvements such as: (1) in the general summarization task (w)AA has not made use of the sentence position and length features; (2) in the query-based summarization, it has not employed any kind of the query processing techniques; (3) another possible enhancement can be reached by introducing the multi-layered graph model that emphasizes the influence of the under sentence and above term-level relations, such as n-grams, phrases, and semantic role arguments levels [20, 21].

References

1. E. Canhasi, I. Kononenko, Multi-document summarization via archetypal analysis of the content-graph joint model. Knowl. Inf. Syst. **41**(3), 821–842 (2014)
2. E. Canhasi, I. Kononenko, Weighted archetypal analysis of the multi-element graph for query-focused multi-document summarization. Expert Syst. Appl. **41**(2), 535–543 (2014)
3. J. Steinberger, K. Ježek, Text summarization and singular value decomposition, *Advances in Information Systems* (Springer, Berlin, 2005), pp. 245–254
4. C.B. Lee, M.S. Kim, H.R. Park, Automatic summarization based on principal component analysis, *Progress in Artificial Intelligence* (Springer, Berlin, 2003), pp. 409–413
5. J. Yeh, Text summarization using a trainable summarizer and latent semantic analysis. Inf. Process. Manag. **41**(1), 75–95 (2005)
6. J.-H. Lee, S. Park, C.M. Ahn, D. Kim, Automatic generic document summarization based on non-negative matrix factorization. Info. Process. Manag. **45**(1), 20–34 (2009)
7. D. Wang, T. Li, S. Zhu, C. Ding, Multi-document summarization via sentence-level semantic analysis and symmetric matrix factorization, in *Proceedings of the 31st Annual International ACM SIGIR Conference on Research and Development in Information Retrieval* (ACM, 2008), pp. 307–314
8. L. Hennig, D. Labor, Topic-based multi-document summarization with probabilistic latent semantic analysis. *Recent Advances in Natural Language Processing (RANLP)* (2009)
9. Y. Ledeneva, R.G. Hernández, R.M. Soto, R.C. Reyes, A. Gelbukh, Em clustering algorithm for automatic text summarization, *Advances in Artificial Intelligence* (Springer, Berlin, 2011), pp. 305–315
10. G. Erkan, D.R. Radev, Lexrank: graph-based lexical centrality as salience in text summarization. J. Artif. Intell. Res. (JAIR) **22**, 457–479 (2004)
11. R. Arora, B. Ravindran, Latent dirichlet allocation and singular value decomposition based multi-document summarization, in: *Eighth IEEE International Conference on Data Mining, ICDM'08* (2008), pp. 713–718

12. S. Park, J.-H. Lee, C.-M. Ahn, J.S. Hong, S.-J. Chun, Query based summarization using non-negative matrix factorization, *Knowledge-Based Intelligent Information and Engineering Systems* (Springer, Berlin, 2006), pp. 84–89
13. J. Otterbacher, G. Erkan, D.R. Radev, Biased lexrank: passage retrieval using random walks with question-based priors. Inf. Process. Manag. **45**(1), 42–54 (2009)
14. C. Bauckhage, C. Thurau, Making archetypal analysis practical, *Pattern Recognition* (Springer, Berlin, 2009), pp. 272–281
15. M. Mørup, L.K. Hansen, Archetypal analysis for machine learning and data mining. Neurocomputing **80**, 54–63 (2012)
16. A. Cutler, L. Breiman, Archetypal analysis. Technometrics **36**(4), 338–347 (1994)
17. M.J. Eugster, F. Leisch, Weighted and robust archetypal analysis. Comput. Stat. Data Anal. **55**(3), 1215–1225 (2011)
18. P. Pentti, T. Unto, Positive matrix factorization: A non-negative factor model with optimal utilization of error estimates of data values. Env. Wiley Online Libr. **5**(2), 111–126 (1994)
19. C.-Y. Lin, Rouge: A package for automatic evaluation of summaries, in *Text Summarization Branches Out: Proceedings of the ACL-04 Workshop* (2004), pp. 74–81
20. A. Khan, N. Salim, Y.J. Kumar, A framework for multi-document abstractive summarization based on semantic role labelling. Appl. Soft Comput. **30**, 737–747 (2015)
21. E. Canhasi, I. Kononenko. Semantic role frames graph-based multidocument summarization, in *Proceedings SiKDD'11* (2011)

Chapter 4
Bounded Matrix Low Rank Approximation

Ramakrishnan Kannan, Mariya Ishteva, Barry Drake and Haesun Park

Abstract Low rank approximation is the problem of finding two matrices $\mathbf{P} \in \mathbb{R}^{m \times k}$ and $\mathbf{Q} \in \mathbb{R}^{k \times n}$ for input matrix $\mathbf{R} \in \mathbb{R}^{m \times n}$, such that $\mathbf{R} \approx \mathbf{PQ}$. It is common in recommender systems rating matrix, where the input matrix $\mathbf{R}$ is bounded in the closed interval $[r_{min}, r_{max}]$ such as $[1, 5]$. In this chapter, we propose a new improved scalable low rank approximation algorithm for such bounded matrices called bounded matrix low rank approximation (BMA) that bounds every element of the approximation $\mathbf{PQ}$. We also present an alternate formulation to bound existing recommender systems algorithms called BALS and discuss its convergence. Our experiments on real-world datasets illustrate that the proposed method BMA outperforms the state-of-the-art algorithms for recommender system such as stochastic gradient descent, alternating least squares with regularization, SVD++ and bias-SVD on real-world datasets such as Jester, Movielens, Book crossing, Online dating, and Netflix.

4.1 Introduction

Matrix low rank approximation of a matrix $\mathbf{R}$ finds matrices $\mathbf{P} \in \mathbb{R}^{n \times k}$ and $\mathbf{Q} \in \mathbb{R}^{k \times m}$ such that $\mathbf{R}$ is well approximated by $\mathbf{PQ}$ i.e., $\mathbf{R} \approx \mathbf{PQ} \in \mathbb{R}^{n \times m}$, where $k < rank(\mathbf{R})$. Low rank approximations vary depending on the constraints imposed on the factors as well as the measure for the difference between $\mathbf{R}$ and $\mathbf{PQ}$. Low rank approximations have produced a huge amount of interest in the data mining and machine learning communities due to its effectiveness for addressing many foundational challenges

R. Kannan (✉) · B. Drake · H. Park
Georgia Institute of Technology, Atlanta, GA, USA
e-mail: rkannan@gatech.edu

B. Drake
e-mail: Barry.Drake@gtri.gatech.edu

H. Park
e-mail: hpark@cc.gatech.edu

M. Ishteva
Vrije Universiteit Brussel (VUB), Brussels, Belgium
e-mail: mariya.ishteva@vub.ac.be

G.R. Naik (ed.), *Non-negative Matrix Factorization Techniques*,
Signals and Communication Technology, DOI 10.1007/978-3-662-48331-2_4

in these application areas. A few prominent techniques of machine learning that use low rank approximation are principal component analysis, factor analysis, latent semantic analysis, and nonnegative matrix factorization (NMF), to name a few.

One of the most important low rank approximations is based on the singular value decompositions (SVD) [8]. Low rank approximation using SVD has many applications over a wide spectrum of disciplines. For example, an image can be compressed by taking the low row rank approximation of its matrix representation using SVD. Similarly, in text data—latent semantic indexing, is a dimensionality reduction technique of a term-document matrix using SVD [5]. The other applications include event detection in streaming data, visualization of a document corpus and many more.

Over the last decade, NMF has emerged as another important low rank approximation technique, where the low rank factor matrices are constrained to have only nonnegative elements. It has received enormous attention and has been successfully applied to a broad range of important problems in areas including, but not limited to, text mining, computer vision, community detection in social networks, visualization, and bioinformatics [13, 20].

In this chapter, we propose a new type of low rank approximation where the elements of the approximation are nonnegative, or more generally, bounded—that is, its elements are within a given range. We call this new low rank approximation bounded matrix low rank approximation (BMA). BMA is different from NMF in that it imposes both upper and lower bounds *on its product* $\mathbf{PQ}$ rather than nonnegativity in each of the low rank factors $\mathbf{P} \geq 0$ and $\mathbf{Q} \geq 0$. Thus, the goal is to obtain a lower rank approximation $\mathbf{PQ}$ of a given input matrix $\mathbf{R}$, where the elements of $\mathbf{PQ}$ and $\mathbf{R}$ are bounded.

Let us consider a numerical example to appreciate the difference between NMF and BMA. Consider the 4×6 matrix with all entries between 1 and 10. In Fig. 4.1, the output low rank approximation is shown between BMA and NMF by running only one iteration for low rank 3. It is important to observe the following.

- All the entries in the BMA are bounded between 1 and 10, whereas, approximation generated out of NMF is not bounded in the same range of input matrix. This difference is hugely pronounced in the case of very large input matrix and the current practice is when an entry in the low rank approximation is beyond the bounds, it is artificially truncated.
- In the case of BMA, as opposed to NMF, each low rank factor is unconstrained and may have negative values.

Also, from this example, it is easy to understand that for certain applications, enforcing constraints on the product of the low rank factors like BMA results in a better approximation error than constraints on individual low rank factors—like NMF.

In order to address the problem of an input matrix with missing elements, we will formulate a BMA that imposes bounds on a low rank matrix that is the best approximate for such matrices. The algorithm design considerations are—(1) Simple implementation (2) Scalable to large data, and (3) Easy parameter tuning with no hyperparameters.

$$\mathbf{R} = \begin{pmatrix} 4 & 2 & 6 & 6 & 1 & 6 \\ 9 & 5 & 9 & 8 & 2 & 9 \\ 2 & 9 & 1 & 6 & 4 & 1 \\ 10 & 8 & 10 & 2 & 9 & 1 \end{pmatrix}$$

Input Bounded Matrix $\mathbf{R} \in [1, 10]$

$$\mathbf{R'}_{BMA} = \begin{pmatrix} 4.832 & 2.233 & 5.118 & 5.824 & 1.000 & 5.878 \\ 8.719 & 4.839 & 9.092 & 8.104 & 2.490 & 9.012 \\ 1.983 & 9.017 & 1.661 & 5.957 & 3.861 & 1.000 \\ 10.000 & 7.896 & 10.000 & 2.106 & 4.871 & 5.599 \end{pmatrix}$$

BMA Output $\mathbf{R}_{BMA}$. The error $\|\mathbf{R} - \mathbf{R'}_{BMA}\|_F^2$ is 40.621.

$$\mathbf{R'}_{NMF} = \begin{pmatrix} 4.939 & 3.378 & 5.312 & 5.021 & 3.131 & 4.736 \\ 8.066 & 5.794 & 8.696 & 8.495 & 4.649 & 8.036 \\ 5.850 & 4.600 & 6.241 & 4.548 & 4.036 & 4.343 \\ 9.693 & 7.549 & \mathbf{10.225} & 5.165 & 8.303 & 4.958 \end{pmatrix}$$

NMF Output $\mathbf{R}_{NMF}$. The error $\|\mathbf{R} - \mathbf{R'}_{NMF}\|_F^2$ is 121.59.

$$\begin{pmatrix} 4.138 & -0.002 & 7.064 \\ 7.037 & -0.008 & 9.429 \\ 3.730 & 0.031 & -4.844 \\ 6.776 & -0.020 & 1.000 \end{pmatrix} \quad \begin{pmatrix} 1.214 & 1.384 & 1.202 & 0.753 & 0.734 & 0.768 \\ -90.835 & 49.650 & -92.455 & 170.147 & -9.311 & -0.525 \\ -0.058 & -0.478 & -0.011 & 0.441 & -0.292 & 0.382 \end{pmatrix}$$

BMA's Left and Right Low Rank Factors $\mathbf{P}_{BMA}$ and $\mathbf{Q}_{BMA}$

$$\begin{pmatrix} 4.021 & 2.806 & 0.153 \\ 8.349 & 4.179 & 0.000 \\ 7.249 & 1.216 & 0.000 \\ 10.015 & 0.725 & 0.467 \end{pmatrix} \quad \begin{pmatrix} 0.727 & 0.605 & 0.770 & 0.431 & 0.557 & 0.416 \\ 0.478 & 0.178 & 0.543 & 1.172 & 0.000 & 1.092 \\ 4.421 & 2.920 & 4.538 & 0.000 & 5.835 & 0.000 \end{pmatrix}$$

NMF's Non-negative Left and Right Low Rank Factors $\mathbf{P}_{NMF}$ and $\mathbf{Q}_{NMF}$

Fig. 4.1 Numerical motivation for BMA

Formally, the BMA problem for an input matrix $\mathbf{R}$ is defined as

$$\begin{aligned} \min_{\mathbf{P},\mathbf{Q}} \quad & \|\mathbf{M} \cdot *(\mathbf{R} - \mathbf{PQ})\|_F^2 \\ \text{subject to} \quad & \\ & r_{min} \leq \mathbf{PQ} \leq r_{max}, \end{aligned} \tag{4.1}$$

where r_{min} and r_{max} are the bounds, $\|\cdot\|_F$ stands for the Frobenius norm and $.*$ is element-wise matrix multiplication. In the case of an input matrix with missing elements, the low rank matrix is approximated only against the known elements of the input matrix. Hence, during error computation the filter matrix **M**, includes only the corresponding elements of the low rank **PQ** for which the values are known. Thus, **M** has '1' everywhere for input matrix **R** with all known elements. However, in the case of a recommender system, the matrix **M** has zero for each of the missing elements of **R**. In fact, for recommender systems, typically only 1 or 2 % of all matrix elements are known.

It should be pointed out that an important application for the above formulation is recommender systems, where the community refers to it as a matrix factorization. The unconstrained version of the above formulation (4.1), was first solved using Stochastic Gradient Descent (SGD) [6] and Alternating Least Squares with Regularization (ALSWR) [31]. However, we have observed that previous research has not leveraged the fact that all the ratings $r_{ij} \in \mathbf{R}$ are bounded within $[r_{min}, r_{max}]$. All existing algorithms *artificially truncate* their final solution to fit within the bounds.

Recently, there have been many innovations introduced into the naive low rank approximation technique such as considering only neighboring entries during the factorization process, time of the ratings, and implicit ratings such as "user watched but did not rate." Hence, it is important to design a bounding framework that seamlessly integrates into the existing sophisticated recommender systems algorithms.

Let $f(\boldsymbol{\Theta}, \mathbf{P}, \mathbf{Q})$ be an existing recommender system algorithm that can predict all the (u, i) ratings, where $\boldsymbol{\Theta} = \{\boldsymbol{\theta}_1, \ldots, \boldsymbol{\theta}_l\}$ is the set of parameters apart from the low rank factors **P**, **Q**. For example, in the recommender system context, certain implicit signals are combined with the explicit ratings such as user watched a movie till the end but did not rate it. We have learned weights for such implicit signals to predict a user's rating. Such weights are represented as parameter $\boldsymbol{\Theta}$. For simplicity, we are slightly abusing the notation here. The $f(\boldsymbol{\Theta}, \mathbf{P}, \mathbf{Q})$ either represents estimating a particular value of (u, i) pair or it represents the complete estimated low rank k matrix $\hat{\mathbf{R}} \in \mathbb{R}^{n \times m}$. The ratings from such recommender system algorithms can be scientifically bounded by the following optimization problem based on low rank approximation to determine the unknown ratings.

$$\begin{aligned} \min_{\boldsymbol{\Theta},\mathbf{P},\mathbf{Q}} \quad & \|\mathbf{M} \cdot *(\mathbf{R} - f(\boldsymbol{\Theta}, \mathbf{P}, \mathbf{Q}))\|_F^2 \\ \text{subject to} \quad & \\ & r_{min} \leq f(\boldsymbol{\Theta}, \mathbf{P}, \mathbf{Q}) \leq r_{max}. \end{aligned} \tag{4.2}$$

Traditionally, regularization is used to control the low rank factors **P** and **Q** from taking larger values. However, this does not guarantee that the value of the product **PQ** is in the given range. We also experimentally show that introducing the bounds on the product of **PQ** outperforms the low rank approximation algorithms with regularization.

In this chapter, we present a survey of current state-of-the-art and foundational material. An explanation of the Block Coordinate Descent (BCD) framework [1] that was used to solve the NMF problem and how it can be extended to solve the Problems (4.1) and (4.2) will be presented. Also, described in this chapter are implementable algorithms and scalable techniques for solving large-scale problems in multicore systems with low memory. Finally, we present substantial experimental results illustrating that the proposed methods outperform the state-of-the-art algorithms for recommender systems such as stochastic gradient descent, alternating least squares with regularization, SVD++, Bias-SVD on real-world datasets such as Jester, Movielens, Book crossing, Online dating, and Netflix. This chapter is based primarily on our earlier work [11, 12]. Notations that are consistent with those in the machine learning literature are used throughout this chapter. A lowercase/uppercase letter such as x or X, is used to denote a scalar; a boldface lowercase letter, such as $\mathbf{x}$, is used to denote a vector; a boldface uppercase letter, such as $\mathbf{X}$, is used to denote a matrix. Indices typically start from 1. When a matrix $\mathbf{X}$ is given, $\mathbf{x}_i$ denotes its ith column, $\mathbf{x}_j^\intercal$ denotes its jth row and x_{ij} or $X(i,j)$ denote its (i,j)th element. For a vector $\mathbf{i}$, $\mathbf{x}(\mathbf{i})$ means that vector $\mathbf{i}$ indexes into the elements of vector $\mathbf{x}$. That is, for $\mathbf{x} = [1, 4, 7, 8, 10]$ and $\mathbf{i} = [1, 3, 5]$, $\mathbf{x}(\mathbf{i}) = [1, 7, 10]$. We have also borrowed certain notations from matrix manipulation scripts such as MATLAB/Octave. For example, the $max(\mathbf{x})$ is the maximal element $x \in \mathbf{x}$ and $max(\mathbf{X})$ is a vector of maximal elements from each column $\mathbf{x} \in \mathbf{X}$.

For the reader's convenience, the notations used in this chapter are summarized in Table 4.1.

Table 4.1 Notations

$\mathbf{R} \in \mathbb{R}^{n \times m}$	Ratings matrix. The missing ratings are indicated by 0, and the given ratings are bounded within $[r_{min}, r_{max}]$
$\mathbf{M} \in \{0, 1\}^{n \times m}$	Indicator matrix. The positions of the missing ratings are indicated by 0, and the positions of the given ratings are indicated by 1
n	Number of users
m	Number of items
k	Value of the reduced rank
$\mathbf{P} \in \mathbb{R}^{n \times k}$	User-feature matrix. Also called as a left low rank factor
$\mathbf{Q} \in \mathbb{R}^{k \times m}$	Feature-item matrix. Also called as a right low rank factor
$\mathbf{p}_x \in \mathbb{R}^{n \times 1}$	xth column vector of $\mathbf{P} = [\mathbf{p_1}, \dots, \mathbf{p_k}]$
$\mathbf{q}_x^\intercal \in \mathbb{R}^{1 \times m}$	xth row vector of $\mathbf{Q} = [\mathbf{q_1}, \dots, \mathbf{q_k}]^\intercal$
$r_{max} > 1$	Maximal rating/upper bound
r_{min}	Minimal rating/lower bound
u	A user
i	An item
$\cdot *$	Element-wise matrix multiplication
$\cdot /$	Element-wise matrix division
$\mathbf{A}(:, i)$	ith column of the matrix $\mathbf{A}$
$\mathbf{A}(i, :)$	ith row of the matrix $\mathbf{A}$
β	Data structure in memory factor
$memsize(v)$	The approximate memory of a variable v (the product of the number of elements in v, the size of each element, and β)
μ	Mean of all known ratings in $\mathbf{R}$
$\mathbf{g} \in \mathbb{R}^n$	Bias of all users u
$\mathbf{h} \in \mathbb{R}^m$	Bias of all items i

4.2 Related Work

This section introduces the BMA and its application to recommender systems and reviews some of the prior research in this area. Following this section is a brief overview of our contributions.

The important milestones in matrix factorization for recommender systems have been achieved due to the Netflix competition (http://www.netflixprize.com/) where the winners were awarded 1 million US Dollars as grand prize.

Funk [6] first proposed matrix factorization for recommender system based on SVD, commonly called the Stochastic Gradient Descent (SGD) algorithm. Paterek [27] improved SGD by combining matrix factorization with baseline estimates. Koren, a member of the winning team of the Netflix prize, improved the results with his remarkable contributions in this area. Koren [17] proposed a baseline estimate based on mean rating, user–movie bias, combined with matrix factorization and called it Bias-SVD. In SVD++ [17], he extended this Bias-SVD with implicit ratings and considered only the relevant neighborhood items during matrix factorization. The Netflix dataset also provided the time of rating. However most of the techniques did not include time in their model. Koren [18] proposed time-svd++, where he extended his previous SVD++ model to include the time information. So far, all matrix factorization techniques discussed here are based on SVD and used gradient descent to solve the problem. Alternatively, Zhou et al. [31] used alternating least squares with regularization (ALSWR). Apart from these directions, there had been other approaches such as Bayesian tensor factorization [29], Bayesian probabilistic modeling [28], graphical modeling of the recommender system problem [24] and weighted low rank approximation with zero weights for the missing values [25]. One of the recent works by Yu et al. [30] also uses coordinate descent to matrix factorization for recommender systems. However, they study the tuning of coordinate descent optimization techniques for a parallel scalable implementation of matrix factorization for recommender systems. A detailed survey and overview of matrix factorization for recommender systems is given in [19].

4.2.1 Our Contributions

Given the above background, we highlight our contributions. We propose a novel matrix factorization called Bounded Matrix Low Rank Approximation (BMA) which imposes a lower and an upper bound for the estimated values of the missing elements in the given matrix. We solve the BMA using the Block Coordinate Descent (BCD) method. From this perspective, this is the first work that uses the block coordinate descent method and experiment BMA for recommender systems. We present the details of the algorithm with supporting technical details and a scalable version of the naive algorithm. It is also important to study imposing bounds for existing recommender systems algorithms. We also propose a novel framework for Bounding existing ALS algorithms (called BALS). Also, we test our BMA algorithm, BALS framework on real-world datasets and compare against state-of-the-art algorithms SGD, SVD++, ALSWR, and Bias-SVD.

4.3 Foundations

In the case of low rank approximation using NMF, the low rank factor matrices are constrained to have only nonnegative elements. However, in the case of BMA, we constrain the elements of their product with an upper and lower bound rather than each of the two low rank factor matrices. In this section, we explain our BCD framework for NMF and subsequently explain using BCD to solve BMA.

4.3.1 NMF and Block Coordinate Descent

Consider a constrained nonlinear optimization problem as follows:

$$\min f(x) \quad \text{subject to } x \in \mathcal{X}, \tag{4.3}$$

where $\mathcal{X}$ is a closed convex subset of $\mathbb{R}^n$. An important assumption to be exploited in the BCD method is that the set $\mathcal{X}$ is represented by a Cartesian product:

$$\mathcal{X} = \mathcal{X}_1 \times \cdots \times \mathcal{X}_l, \tag{4.4}$$

where $\mathcal{X}_j, j = 1, \ldots, l$, is a closed convex subset of $\mathbb{R}^{N_j}$, satisfying $n = \sum_{j=1}^{l} N_j$. Accordingly, vector $\mathbf{x}$ is partitioned as $\mathbf{x} = (\mathbf{x}_1, \ldots, \mathbf{x}_l)$ so that $\mathbf{x}_j \in \mathcal{X}_j$ for $j = 1, \ldots, l$. The BCD method solves for $\mathbf{x}_j$ fixing all other subvectors of $\mathbf{x}$ in a cyclic manner. That is, if $\mathbf{x}^{(i)} = (\mathbf{x}_1^{(i)}, \ldots, \mathbf{x}_l^{(i)})$ is given as the current iterate at the ith step, the algorithm generates the next iterate $\mathbf{x}^{(i+1)} = (\mathbf{x}_1^{(i+1)}, \ldots, \mathbf{x}_l^{(i+1)})$ block by block, according to the solution of the following subproblem:

$$\mathbf{x}_j^{(i+1)} \leftarrow \operatorname*{argmin}_{\xi \in \mathcal{X}_j} f(\mathbf{x}_1^{(i+1)}, \ldots, \mathbf{x}_{j-1}^{(i+1)}, \xi, \mathbf{x}_{j+1}^{(i)}, \ldots, \mathbf{x}_l^{(i)}). \tag{4.5}$$

Also known as a *nonlinear Gauss–Seidel* method [1], this algorithm updates one block each time, always using the most recently updated values of other blocks $\mathbf{x}_{\tilde{j}}, \tilde{j} \neq j$. This is important since it ensures that after each update the objective function value does not increase. For a sequence $\left\{\mathbf{x}^{(i)}\right\}$ where each $\mathbf{x}^{(i)}$ is generated by the BCD method, the following property holds.

Theorem 1 *Suppose* f *is continuously differentiable in* $\mathcal{X} = \mathcal{X}_1 \times \cdots \times \mathcal{X}_l$, *where* $\mathcal{X}_j$, $j = 1, \ldots, l$, *are closed convex sets. Furthermore, suppose that for all* j *and* i, *the minimum of*

$$\min_{\xi \in \mathcal{X}_j} f(\mathbf{x}_1^{(i+1)}, \ldots, \mathbf{x}_{j-1}^{(i+1)}, \xi, \mathbf{x}_{j+1}^{(i)}, \ldots, \mathbf{x}_l^{(i)})$$

is uniquely attained. Let $\{\mathbf{x}^{(i)}\}$ *be the sequence generated by the block coordinate descent method as in Eq. (4.5). Then, every limit point of* $\{\mathbf{x}^{(i)}\}$ *is a stationary point. The uniqueness of the minimum is not required when* $l = 2$ *[9].*

The proof of this theorem for an arbitrary number of blocks is shown in Bertsekas [1]. For a nonconvex optimization problem, often we can expect the stationarity of a limit point [22] from a good algorithm.

When applying the BCD method to a constrained nonlinear programming problem, it is critical to wisely choose a partition of $\mathcal{X}$, whose Cartesian product constitutes $\mathcal{X}$. An important criterion is whether the subproblems in Eq. (4.5) are efficiently solvable: for example, if the solutions of subproblems appear in closed form, each update can be computed fast. In addition, it is worth checking how the solutions of subproblems depend on each other. The BCD method requires that the most recent values be used for each subproblem in Eq. (4.5). When the solutions of subproblems depend on each other, they have to be computed sequentially to make use of the most recent values; if solutions for some blocks are independent from each other, however, they can be computed simultaneously. We discuss how different choices of partitions lead to different NMF algorithms. Three cases of partitions are discussed below.

4.3.1.1 BCD with Two Matrix Blocks—ANLS Method

For convenience, we first assume all the elements of the input matrix are known and hence we ignore $\mathbf{M}$. The most natural partitioning of the variables is to have two large blocks, $\mathbf{P}$ and $\mathbf{Q}$. In this case, following the BCD method in Eq. (4.5), we alternate solving

$$\mathbf{P} \leftarrow \arg\min_{\mathbf{P} \geq 0} f(\mathbf{P}, \mathbf{Q}) \quad \text{and} \quad \mathbf{Q} \leftarrow \arg\min_{\mathbf{Q} \geq 0} f(\mathbf{P}, \mathbf{Q}). \tag{4.6}$$

Since the subproblems are nonnegativity constrained least squares (NLS) problems, the two-block BCD method has been called the alternating nonnegative least square (ANLS) framework [14, 16, 22].

4.3.1.2 BCD with 2*k* Vector Blocks—HALS/RRI Method

Let us now partition the unknowns into $2k$ blocks in which each block is a column of $\mathbf{P}$ or a row of $\mathbf{Q}$, as explained in Fig. 4.2. In this case, it is easier to consider the objective function in the following form:

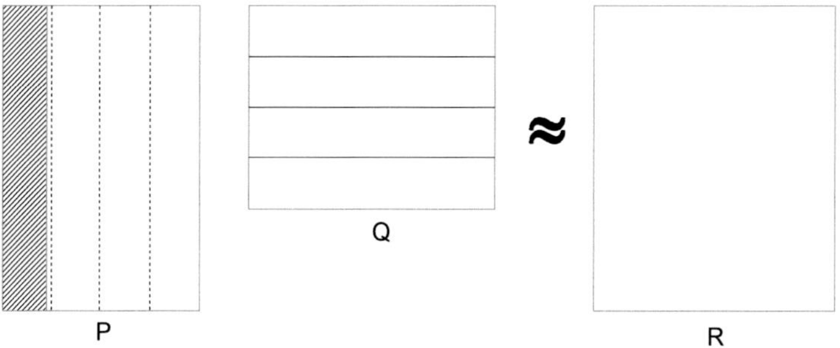

Fig. 4.2 BCD with $2k$ vector blocks

$$f(\mathbf{p}_1, \ldots, \mathbf{p}_k, \mathbf{q}_1^{\mathsf{T}}, \ldots, \mathbf{q}_k^{\mathsf{T}}) = \|\mathbf{R} - \sum_{j=1}^{k} \mathbf{p}_j \mathbf{q}_j^T \|_F^2, \tag{4.7}$$

where $\mathbf{P} = [\mathbf{p}_1, \ldots, \mathbf{p}_k] \in \mathbb{R}_+^{n \times k}$ and $\mathbf{Q} = [\mathbf{q}_1, \ldots, \mathbf{q}_k]^{\mathsf{T}} \in \mathbb{R}_+^{k \times m}$. The form in Eq. (4.7) represents that $\mathbf{R}$ is approximated by the sum of k rank-one matrices.

Following the BCD scheme, we can minimize f by iteratively solving

$$\mathbf{p}_i \leftarrow \arg\min_{\mathbf{p}_i \geq 0} f(\mathbf{p}_1, \ldots, \mathbf{p}_k, \mathbf{q}_1^{\mathsf{T}}, \ldots, \mathbf{q}_k^{\mathsf{T}})$$

for $i = 1, \ldots, k$, and

$$\mathbf{q}_i^{\mathsf{T}} \leftarrow \arg\min_{\mathbf{q}_i^{\mathsf{T}} \geq 0} f(\mathbf{p}_1, \ldots, \mathbf{p}_k, \mathbf{q}_1^{\mathsf{T}}, \ldots, \mathbf{q}_k^{\mathsf{T}})$$

for $i = 1, \ldots, k$.

The $2k$-block BCD algorithm has been studied as Hierarchical Alternating Least Squares (HALS) proposed by Cichocki et al. [3, 4] and independently by Ho et al. [10] as rank-one residue iteration (RRI).

4.3.1.3 BCD with $k(n+m)$ Scalar Blocks

We can also partition the variables with the smallest $k(n+m)$ element blocks of scalars as in Fig. 4.3, where every element of $\mathbf{P}$ and $\mathbf{Q}$ is considered as a block in the context of Theorem 1. To this end, it helps to write the objective function as a quadratic function of scalar p_{ij} or q_{ij} assuming all other elements in $\mathbf{P}$ and $\mathbf{Q}$ are fixed:

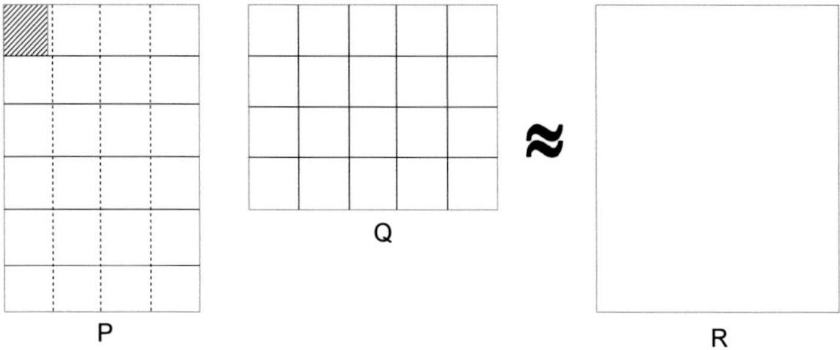

Fig. 4.3 BCD with $k(n+m)$ scalar blocks

$$f(p_{ij}) = \| \left(\mathbf{r}_i^\mathsf{T} - \sum_{\tilde{k} \neq j} p_{i\tilde{k}} \mathbf{q}_{\tilde{k}}^\mathsf{T} \right) - p_{ij} \mathbf{q}_j^\mathsf{T} \|_2^2 + \text{const.}, \tag{4.8a}$$

$$f(q_{ij}) = \| \left(\mathbf{r}_j - \sum_{\tilde{k} \neq i} \mathbf{p}_{\tilde{k}} q_{\tilde{k}j} \right) - \mathbf{p}_i q_{ij} \|_2^2 + \text{const.}, \tag{4.8b}$$

where $\mathbf{r}_i^\mathsf{T}$ and $\mathbf{r}_j$ denote the ith row and the jth column of $\mathbf{R}$, respectively.

Kim et al. [15] discuss solving the NMF problem using the BCD method.

4.3.2 Bounded Matrix Low Rank Approximation

The building blocks of BMA are column vectors $\mathbf{p}_x$ and row vectors $\mathbf{q}_x^\mathsf{T}$ of the matrix $\mathbf{P}$ and $\mathbf{Q}$ respectively. In this section, we discuss the idea behind finding these vectors $\mathbf{p}_x$ and $\mathbf{q}_x^\mathsf{T}$ such that all the elements in $\mathbf{T} + \mathbf{p}_x\mathbf{q}_x^\mathsf{T} \in [r_{min}, r_{max}]$ and the error $\|\mathbf{M} \cdot *(\mathbf{R} - \mathbf{PQ})\|_F^2$ is reduced. Here, $\mathbf{T} = \sum_{j=1, j\neq x}^{k} \mathbf{p}_j\mathbf{q}_j^\mathsf{T}$.

Problem (4.1) can be equivalently represented with a set of rank-one matrices $\mathbf{p}_x\mathbf{q}_x^\mathsf{T}$ as

$$\begin{aligned} \min_{\mathbf{p}_x, \mathbf{q}_x} \quad & \|\mathbf{M} \cdot *(\mathbf{R} - \mathbf{T} - \mathbf{p}_x\mathbf{q}_x^\mathsf{T})\|_F^2 \\ & \qquad \forall x = [1, k] \\ \text{subject to} \quad & \\ & \mathbf{T} + \mathbf{p}_x\mathbf{q}_x^\mathsf{T} \leq r_{max} \\ & \mathbf{T} + \mathbf{p}_x\mathbf{q}_x^\mathsf{T} \geq r_{min} \end{aligned} \tag{4.9}$$

Thus, we take turns solving for $\mathbf{p}_x$ and $\mathbf{q}_x^\mathsf{T}$. That is, assume we know $\mathbf{p}_x$ and find $\mathbf{q}_x^\mathsf{T}$ and vice versa. In the entire section, we assume fixing column $\mathbf{p}_x$ and finding row $\mathbf{q}_x^\mathsf{T}$. Without loss of generality, all the discussions pertaining to finding $\mathbf{q}_x^\mathsf{T}$ with fixed $\mathbf{p}_x$ hold for the other scenario of finding $\mathbf{p}_x$ with fixed $\mathbf{q}_x^\mathsf{T}$.

There are different orders of updates of vector blocks when solving Problem (4.9). For example,

$$\mathbf{p}_1 \rightarrow \mathbf{q}_1^{\mathsf{T}} \rightarrow \cdots \rightarrow \mathbf{p}_k \rightarrow \mathbf{q}_k^{\mathsf{T}} \tag{4.10}$$

and

$$\mathbf{p}_1 \rightarrow \cdots \rightarrow \mathbf{p}_k \rightarrow \mathbf{q}_1^{\mathsf{T}} \rightarrow \cdots \rightarrow \mathbf{q}_k^{\mathsf{T}}. \tag{4.11}$$

Kim et al. [15] prove that Eq. (4.7) satisfies the formulation of BCD method. Equation (4.7) when extended with the matrix $\mathbf{M}$ becomes Eq. (4.9). Here, the matrix $\mathbf{M}$ is like a filter matrix that defines the elements of $(\mathbf{R}-\mathbf{T}-\mathbf{p}_x\mathbf{q}_x^{\mathsf{T}})$ to be included for the norm computation. Thus, Problem (4.9) is similar to Problem (4.7) and we can solve by applying $2k$ block BCD to update $\mathbf{p}_x$ and $\mathbf{q}_x^{\mathsf{T}}$ iteratively, although Eq. (4.9) appears not to satisfy the BCD requirements directly. We focus on the scalar block case, as it is convenient to explain regarding imposing bounds on the product of the low rank factors $\mathbf{PQ}$.

Also, according to BCD, the independent elements in a block can be computed simultaneously. Here, the computations of the elements $q_{xi}, q_{xj} \in \mathbf{q}_x^{\mathsf{T}}, i \neq j$, are independent of each other. Hence, the problem of finding row $\mathbf{q}_x^{\mathsf{T}}$ fixing column $\mathbf{p}_x$ is equivalent to solving the following problem

$$\begin{aligned} &\min_{q_{xi}} \|\mathbf{M}(:,i) \cdot *((\mathbf{R}-\mathbf{T})(:,i) - \mathbf{p}_x q_{xi})\|_F^2 \\ &\forall i = [1,m],\ \forall x = [1,k] \\ \text{subject to}\quad & \\ &\mathbf{T}(:,i) + \mathbf{p}_x q_{xi} \leq r_{max} \\ &\mathbf{T}(:,i) + \mathbf{p}_x q_{xi} \geq r_{min} \end{aligned} \tag{4.12}$$

To construct the row vector $\mathbf{q}_x^{\mathsf{T}}$, we use $k(n+m)$ scalar blocks based on problem formulation (4.12). Theorem 3 identifies these best elements that construct $\mathbf{q}_x^{\mathsf{T}}$. As shown in Fig. 4.4, given the **bold** blocks, $\mathbf{T}$, $\mathbf{R}$, and $\mathbf{p}_x$, we find the row vector $\mathbf{q}_x^{\mathsf{T}} = [q_{x1}, q_{x2}, \ldots, q_{xm}]$ for Problem (4.12). For this, let us understand the boundary values of q_{xi} by defining two vectors, $\mathbf{l}$ bounding q_{xi} from below, and $\mathbf{u}$ bounding q_{xi} from above, i.e., $max(\mathbf{l}) \leq q_{xi} \leq min(\mathbf{u})$.

Definition 1 The lower bound vector $\mathbf{l} = [l_1, \ldots, l_n] \in \mathbb{R}^n$ and the upper bound vector $\mathbf{u} = [u_1, \ldots, u_n] \in \mathbb{R}^n$ for a given $\mathbf{p}_x$ and $\mathbf{T}$ that bound q_{xi} are defined $\forall j \in [1,n]$ as

$$l_j = \begin{cases} \dfrac{r_{min} - \mathbf{T}(j,i)}{p_{jx}}, & p_{jx} > 0 \\ \dfrac{r_{max} - \mathbf{T}(j,i)}{p_{jx}}, & p_{jx} < 0 \\ -\infty, & \text{otherwise} \end{cases}$$

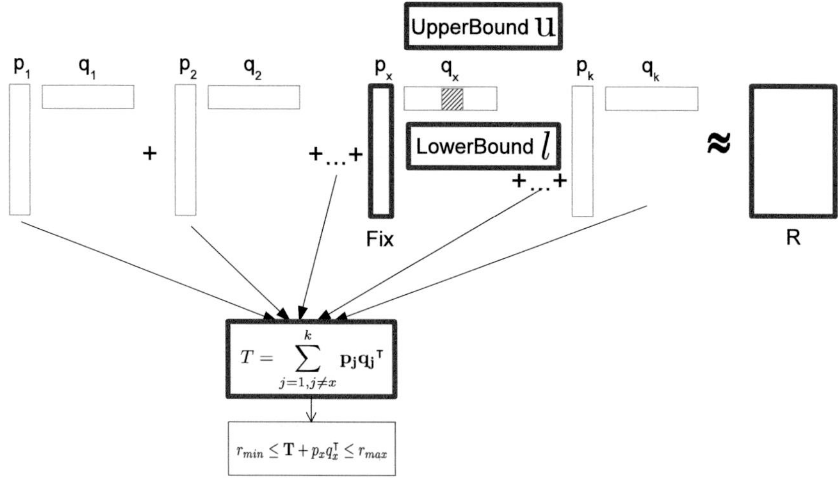

Fig. 4.4 Bounded Matrix Low Rank Approximation solution overview

and

$$u_j = \begin{cases} \dfrac{r_{max} - \mathbf{T}(j,i)}{p_{jx}}, & p_{jx} > 0 \\ \dfrac{r_{min} - \mathbf{T}(j,i)}{p_{jx}}, & p_{jx} < 0 \\ \infty, & \text{otherwise.} \end{cases}$$

It is important to observe that the defined $\mathbf{l}$ and $\mathbf{u}$—referred as `LowerBounds` and `UpperBounds` in Algorithm 1, are for a given $\mathbf{p}_x$ and $\mathbf{T}$ to bound q_{xi}. Alternatively, if we are solving $\mathbf{p}_x$ for a given $\mathbf{T}$ and $\mathbf{q}_x$, the above function correspondingly represents the possible lower and upper bounds for p_{ix}, where $\mathbf{l}, \mathbf{u} \in \mathbb{R}^m$.

Theorem 2 *Given* $\mathbf{R}$, $\mathbf{T}$, $\mathbf{p}_x$, *the* q_{xi} *is always bounded as* $max(\mathbf{l}) \leq q_{xi} \leq min(\mathbf{u})$.

Proof It is easy to see that if $q_{xi} < max(\mathbf{l})$ or $q_{xi} > min(\mathbf{u})$, then $\mathbf{T}(:,i) + \mathbf{p}_x q_{xi}^\mathsf{T} \notin [r_{min}, r_{max}]$. □

Here, it is imperative to note that if q_{xi}, results in $\mathbf{T}(:,i) + \mathbf{p}_x q_{xi}^\mathsf{T} \notin [r_{min}, r_{max}]$, this implies that q_{xi} is either less than the $max(\mathbf{l})$ or greater than the $min(\mathbf{u})$. It cannot be any other inequality.

Given the boundary values of q_{xi}, Theorem 3 defines the solution to Problem (4.12).

Theorem 3 *Given* $\mathbf{T}$, $\mathbf{R}$, $\mathbf{p}_x$, $\mathbf{l}$ *and* $\mathbf{u}$, *let*

$$\hat{q_{xi}} = ([\mathbf{M}(:,i) \cdot *(\mathbf{R} - \mathbf{T})(:,i)]^\mathsf{T} \mathbf{p}_x)/(\|\mathbf{M}(:,i) \cdot *\mathbf{p}_x\|_2^2).$$

The unique solution q_{xi}*—referred as* `FindElement` *in Algorithm 1 to the least squares problem* (4.12) *is given as*

$$q_{xi} = \begin{cases} max(\mathbf{l}), & \text{if } \hat{q_{xi}} < max(\mathbf{l}) \\ min(\mathbf{u}), & \text{if } \hat{q_{xi}} > min(\mathbf{u}) \\ \hat{q_{xi}}, & \text{otherwise.} \end{cases}$$

Proof Out of Boundary: $q_{xi} < max(\mathbf{l})$ or $q_{xi} > min(\mathbf{u})$. Under this circumstance, the best value for q_{xi} is either $max(\mathbf{l})$ or $min(\mathbf{u})$. We can prove this by contradiction. Let us assume there exists a $\tilde{q_{xi}} = max(\mathbf{l}) + \delta$; $\delta > 0$ that is optimal to the Problem (4.12) for $q_{xi} < max(\mathbf{l})$. However, for $q_{xi} = max(\mathbf{l}) < \tilde{q_{xi}}$ is still a feasible solution for the Problem (4.12). Also, there does not exist a feasible solution that is less than $max(\mathbf{l})$, because the Problem (4.12) is quadratic in q_{xi}. Hence for $q_{xi} < max(\mathbf{l})$, the optimal value for the Problem (4.12) is $max(\mathbf{l})$. In similar direction we can show that the optimal value of q_{xi} is $min(\mathbf{u})$ for $q_{xi} > min(\mathbf{u})$.
Within Boundary: $max(\mathbf{l}) \le q_{xi} \le min(\mathbf{u})$.

Let us consider the objective function of the unconstrained optimization problem (4.12). That is, $f = \min_{q_{xi}} \|\mathbf{M}(:,i) \cdot *((\mathbf{R}-\mathbf{T})(:,i) - \mathbf{p}_x q_{xi})\|_2^2$. The minimum value is determined by taking the derivative of f with respect to q_{xi} and equating it to zero.

$$\begin{aligned}
\frac{\partial f}{\partial q_{xi}} &= \frac{\partial}{\partial q_{xi}} \left(\sum_{\substack{\texttt{all known ratings} \\ \texttt{in column } i}} (\mathbf{E}_i - \mathbf{p}_x q_{xi})^2 \right) \quad (\texttt{where } \mathbf{E} = \mathbf{R} - \mathbf{T}) \\
&= \frac{\partial}{\partial q_{xi}} \left(\sum_{\substack{\texttt{all known ratings} \\ \texttt{in column } i}} (\mathbf{E}_i - \mathbf{p}_x q_{xi})^\mathsf{T} (\mathbf{E}_i - \mathbf{p}_x q_{xi}) \right) \\
&= \frac{\partial}{\partial q_{xi}} \left(\sum_{\substack{\texttt{all known ratings} \\ \texttt{in column } i}} (\mathbf{E}_i^\mathsf{T} - q_{xi}\mathbf{p}_x^\mathsf{T})(\mathbf{E}_i - \mathbf{p}_x q_{xi}) \right) \\
&= \frac{\partial}{\partial q_{xi}} \left(\sum_{\substack{\texttt{all known ratings} \\ \texttt{in column } i}} q_{xi}^2 \mathbf{p}_x^\mathsf{T}\mathbf{p}_x - q_{xi}\mathbf{E}_i^\mathsf{T}\mathbf{p}_x - q_{xi}\mathbf{p}_x^\mathsf{T}\mathbf{E}_i + \mathbf{E}_i^\mathsf{T}\mathbf{E}_i \right) \\
&= 2\|\mathbf{M}(:,i). * \mathbf{p}_x\|_2^2 q_{xi} - 2[\mathbf{M}(:,i). * (\mathbf{R}-\mathbf{T})(:,i)]^\mathsf{T}\mathbf{p}_x
\end{aligned} \tag{4.13}$$

Now, equating $\frac{\partial f}{\partial q_{xi}}$ to zero will yield the optimum solution for the unconstrained optimization problem (4.12) as $q_{xi} = ([\mathbf{M}(:,i) \cdot *(\mathbf{R}-\mathbf{T})(:,i)]^\mathsf{T}\mathbf{p}_x)/(\|\mathbf{M}(:,i) \cdot *\mathbf{p}_x\|_2^2)$. □

In a similar way the proof for Theorem 4 can also be established.

4.3.3 Bounding Existing ALS Algorithms (BALS)

Over the last few years, recommender system algorithms have improved by leaps and bounds. The additional sophistication such as using only nearest neighbors during factorization, implicit ratings, time, etc., gave only a diminishing advantage for the Root Mean Square Error (RMSE) scores. That is, the improvement in RMSE score over the naive low rank approximation with implicit ratings is more than the improvement attained by utilizing both implicit ratings and time. Today, these algorithms artificially truncate the estimated unknown ratings. However, it is important to investigate establishing bounds scientifically on these existing alternating least squares (ALS) type algorithms.

Using matrix block BCD, we introduce a temporary variable $\mathbf{Z} \in \mathbb{R}^{n \times m}$ with box constraints to solve Problem (4.1),

$$\begin{aligned} &\min_{\boldsymbol{\Theta},\mathbf{Z},\mathbf{P},\mathbf{Q}} \|\mathbf{M} \cdot *(\mathbf{R} - \mathbf{Z})\|_F^2 + \alpha\|\mathbf{Z} - f(\boldsymbol{\Theta}, \mathbf{P}, \mathbf{Q})\|_F^2 \\ &\text{subject to} \\ &\qquad r_{min} \leq \mathbf{Z} \leq r_{max}. \end{aligned} \tag{4.14}$$

The key question is identifying optimal $\mathbf{Z}$. We assume the iterative algorithm has a specific update order, for example, $\boldsymbol{\theta}_1 \rightarrow \cdots \rightarrow \boldsymbol{\theta}_l \rightarrow \mathbf{P} \rightarrow \mathbf{Q}$. Before updating these parameters, we should have an optimal $\mathbf{Z}$, with the most recent values of $\boldsymbol{\Theta}$, $\mathbf{P}$, $\mathbf{Q}$.

Theorem 4 *The optimal* $\mathbf{Z}$, *given* $\mathbf{R}, \mathbf{P}, \mathbf{Q}, \mathbf{M}, \boldsymbol{\Theta}$, *is* $\frac{\mathbf{M}.*(\mathbf{R}+\alpha f(\boldsymbol{\Theta},\mathbf{P},\mathbf{Q}))}{1+\alpha} + \mathbf{M}'.*f(\boldsymbol{\Theta}), \mathbf{P}, \mathbf{Q})$, *where* $\mathbf{M}'$ *is the complement of the indicator boolean matrix* $\mathbf{M}$.

Proof Given $\mathbf{R}, \mathbf{P}, \mathbf{Q}, \mathbf{M}, \boldsymbol{\Theta}$, the optimal $\mathbf{Z}$, is obtained by solving the following optimization problem.

$$\begin{aligned} &G = \min_{\mathbf{Z}} \|\mathbf{M} \cdot *(\mathbf{R} - \mathbf{Z})\|_F^2 + \alpha\|\mathbf{Z} - f(\boldsymbol{\Theta}, \mathbf{P}, \mathbf{Q})\|_F^2 \\ &\text{subject to} \\ &\qquad r_{min} \leq \mathbf{Z} \leq r_{max}. \end{aligned} \tag{4.15}$$

Taking the gradient $\frac{\partial G}{\partial Z}$ of the above equation to find the optimal solution yields, $\frac{\mathbf{M}.*(\mathbf{R}+\alpha f(\boldsymbol{\Theta},\mathbf{P},\mathbf{Q}))}{1+\alpha} + \mathbf{M}'.*f(\boldsymbol{\Theta}, \mathbf{P}, \mathbf{Q})$. The derivation proceeds in the same way as explained in Eq. (4.13). □

Similar to the proof of Theorem 3, we can show that if the values of $\mathbf{Z}$ are outside $[r_{min}, r_{max}]$, that is, $\mathbf{Z} > r_{max}$ and $\mathbf{Z} < r_{min}$, the optimal value is $\mathbf{Z} = r_{max}$ and $\mathbf{Z} = r_{min}$ respectively.

In the next section, the implementation of the algorithm for *BMA* and its variants such as scalable and block implementations will be studied. Also, imposing bounds on existing ALS algorithms using *BALS* will be investigated. As an example, we will take existing algorithms from Graphchi [21] implementations and study imposing bounds using the *BALS* framework.

4.4 Implementations

4.4.1 Bounded Matrix Low Rank Approximation

Given the discussion in the previous sections, we now have the necessary tools to construct the algorithm. In Algorithm 1, the **l** and **u** from Theorem 2 are referred to as `LowerBounds` and `UpperBounds`, respectively. Also, q_{xi} from Theorem 3 is referred to as `FindElement`. The BMA algorithm has three major functions: (1) Initialization, (2) Stopping Criteria, and (3) Find the low rank factors **P** and **Q**. In later sections, the initialization and stopping criteria are explained in detail. For now, we assume that two initial matrices **P** and **Q** are required, such that $\mathbf{PQ} \in [r_{min}, r_{max}]$, and that a stopping criterion will be used for terminating the algorithm, when the constructed matrices **P** and **Q** provide a good representation of the given matrix **R**.

In the case of the BMA algorithm, since multiple elements can be updated independently, we reorganize the scalar block BCD into $2k$ vector blocks. The BMA algorithm is presented as Algorithm 1.

Algorithm 1 works very well and yields low rank factors **P** and **Q** for a given matrix **R** such that $\mathbf{PQ} \in [r_{min}, r_{max}]$. However, when applied for very large scale matrices, such as recommender systems, it can only be run on machines with a large amount of memory. We address scaling the algorithm on multi-core systems and machines with low memory in the next section.

4.4.2 Scaling up Bounded Matrix Low Rank Approximation

In this section, we address the issue of scaling the algorithm for large matrices with missing elements. Two important aspects of making the algorithm run for large matrices are running time and memory. We discuss the parallel implementation of the algorithm, which we refer to as *Parallel Bounded Matrix Low Rank Approximation*. Subsequently, we also discuss a method called *Block Bounded Matrix Low Rank Approximation*, which will outline the details of executing the algorithm for large matrices in low memory systems. Let us start this section by discussing *Parallel Bounded Matrix Low Rank Approximation*.

input : Matrix $\mathbf{R} \in \mathbb{R}^{n \times m}$, $r_{min}, r_{max} > 1$, reduced rank k
output: Matrix $\mathbf{P} \in \mathbb{R}^{n \times k}$ and $\mathbf{Q} \in \mathbb{R}^{k \times m}$

// Rand initialization of $\mathbf{P}$, $\mathbf{Q}$.
Initialize $\mathbf{P}$, $\mathbf{Q}$ *as nonnegative random matrices* ;
// modify random $\mathbf{PQ}$ such that
// $\mathbf{PQ} \in [r_{min}, r_{max}]$
// maxelement of $\mathbf{PQ}$ without first
// column of $\mathbf{P}$ and first row of $\mathbf{Q}$
$maxElement = max(\mathbf{P}(:, 2 : end) * \mathbf{Q}(2 : end, :))$;
$\alpha \leftarrow \sqrt{\frac{r_{max} - 1}{maxElement}}$;
$\mathbf{P} \leftarrow \alpha \cdot \mathbf{P}$;
$\mathbf{Q} \leftarrow \alpha \cdot \mathbf{Q}$;
$\mathbf{P}(:, 1) \leftarrow 1$;
$\mathbf{Q}(1, :) \leftarrow 1$;
$\mathbf{M} \leftarrow$ `ComputeRatedBinaryMatrix`($\mathbf{R}$);
while *stopping criteria not met* **do**
  for $x \leftarrow 1$ **to** k **do**
    $\mathbf{T} \leftarrow \sum_{j=1, j \neq x}^{k} \mathbf{p}_j \mathbf{q}_j^{\mathsf{T}}$;
    for $i \leftarrow 1$ **to** m **do**
      // Find vectors $\mathbf{l}, \mathbf{u} \in \mathbb{R}^n$ as in Definition 1
      $\mathbf{l} \leftarrow$ `LowerBounds`($r_{min}, r_{max}, \mathbf{T}, i, \mathbf{p}_x$);
      $\mathbf{u} \leftarrow$ `UpperBounds`($r_{min}, r_{max}, \mathbf{T}, i, \mathbf{p}_x$);
      // Find vector $\mathbf{q}_x^{\mathsf{T}}$ fixing $\mathbf{p}_x$ as in Theorem 3
      $q_{xi} \leftarrow$ `FindElement`($\mathbf{p_x}, \mathbf{M}, \mathbf{R}, \mathbf{T}, i, x$);
      if $q_{xi} < max(\mathbf{l})$ **then**
        $q_{xi} \leftarrow max(\mathbf{l})$;
      else if $q_{xi} > min(\mathbf{u})$ **then**
        $q_{xi} \leftarrow min(\mathbf{u})$;
    for $i \leftarrow 1$ **to** n **do**
      // Find vectors $\mathbf{l}, \mathbf{u} \in \mathbb{R}^m$ as in Definition 1
      $\mathbf{l} \leftarrow$ `LowerBounds`($r_{min}, r_{max}, \mathbf{T}, i, \mathbf{q}_x^{\mathsf{T}}$);
      $\mathbf{u} \leftarrow$ `UpperBounds`($r_{min}, r_{max}, \mathbf{T}, i, \mathbf{q}_x^{\mathsf{T}}$);
      // Find vector $\mathbf{p}_x$ fixing $\mathbf{q}_x^{\mathsf{T}}$ as in Theorem 3
      $p_{ix} \leftarrow$ `FindElement`($\mathbf{q}_x^{\mathsf{T}}, \mathbf{M}^{\mathsf{T}}, \mathbf{R}^{\mathsf{T}}, \mathbf{T}^{\mathsf{T}}, i, x$);
      if $p_{ix} < max(\mathbf{l})$ **then**
        $p_{ix} \leftarrow max(\mathbf{l})$;
      else if $p_{ix} > min(\mathbf{u})$ **then**
        $p_{ix} \leftarrow min(\mathbf{u})$;

Algorithm 1: Bounded Matrix Low Rank Approximation (BMA)

4.4.2.1 Parallel Bounded Matrix Low Rank Approximation

In the case of the BCD method, the solutions of the subproblems that depend on each other have to be computed sequentially to make use of the most recent values. However, if solutions for some blocks are independent of each other, it is possible to compute them simultaneously. We can observe that, according to Theorem 3, all elements $q_{xi}, q_{xj} \in \mathbf{q}_x^{\mathsf{T}}, i \neq j$ are independent of each other. We are leveraging this characteristic to parallelize the **for** loops in Algorithm 1. Nowadays, virtually all commercial processors have multiple cores. Hence, we can parallelize finding the q_{xi}'s across multiple cores. Since it is trivial to change the **for** in step 12 and step 20 of Algorithm 1 to **parallel for** the details will be omitted.

It is obvious to see that the **T** at step 11 in Algorithm 1 requires the largest amount of memory. Also, the function *FindElement* in step 15 takes a sizable amount of memory. Hence, it is not possible to run the algorithm for large matrices on machines with low memory, e.g., with rows and columns on the scale of 100,000's. Thus, we propose the following algorithm to mitigate this limitation: Block BMA.

4.4.2.2 Block Bounded Matrix Low Rank Approximation

To facilitate understanding of this section, let us define β—a data structure in memory factor. That is, maintaining a floating scalar as a sparse matrix with one element or full matrix with one element takes different amounts of memory. This is because of the data structure that is used to represent the numbers in the memory. The amount of memory is also dependent on using single or double precision floating point precision. Typically, in MATLAB, the data structure in memory factor β for a full matrix is around 10. Similarly, in Java, the β factor for maintaining a number in an ArrayList is around 8. Let, $memsize(v)$ be the function that returns the approximate memory size of a variable v. Generally, $memsize(v)$ = number of elements in v * size of each element * β. Consider an example of maintaining 1000 floating point numbers on an ArrayList of a Java program. The approximate memory would be 1000*4*8 = 32,000 bytes $\approx$ 32 KB in contrast to the actual 4 KB due to the factor $\beta = 8$.

As discussed earlier, for most of the real-world large datasets such as Netflix, Yahoo music, online dating, book crossing, etc., it is impossible to keep the entire matrix **T** in memory. Also, notice that, according to Theorem 3 and Definition 1, we need only the ith column of **T** to compute q_{xi}. The block size of $q_{x\mathrm{i}}$ to compute in one core of the machine is dependent on the size of **T** and *FindElements* that fits in memory.

On the one hand, partition $\mathbf{q}_x$ to fit the maximum possible **T** and *FindElements* in the entire memory of the system. If very small partitions are created such that, we can give every core some amount of work so that the processing capacity of the system is not underutilized. The disadvantage of the former, is that only one core is used. However, in the latter case, there is a significant communication overhead. Figure 4.5 gives the pictorial view of the Block Bounded Matrix Low Rank Approximation.

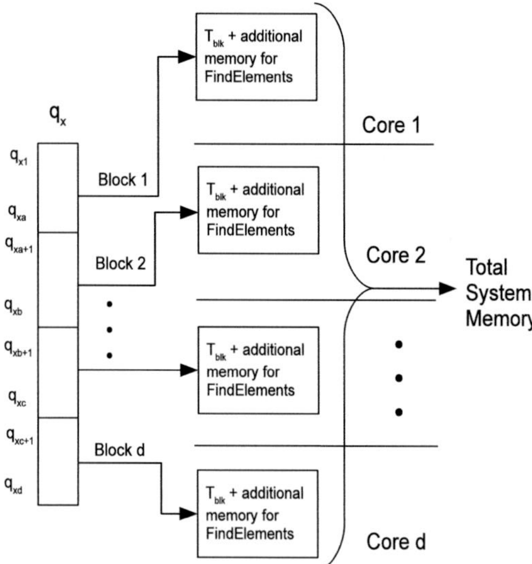

Fig. 4.5 Block bounded matrix low rank approximation

We determined the number of blocks = memsize(full(**R**) + other variables of FindElement)/(system memory $*$ number of d cores). The full(**R**) is a nonsparse representation and $d \leq$ number of cores available in the system. Typically, for most of the datasets, we achieved minimum running time when we used 4 cores and 16 blocks. That is, we find 1/16-th of $\mathbf{q}_x^\mathsf{T}$ concurrently on 4 cores.

For convenience, we have presented the Block BMA as Algorithm 2. We describe only the algorithm to find the partial vector of $\mathbf{q}_x^\mathsf{T}$ given $\mathbf{p}_x$. To find more than one element, Algorithm 1 is modified such that the vectors $\mathbf{l}$, $\mathbf{u}$, $\mathbf{p}_x$ are matrices $\mathbf{L}$, $\mathbf{U}$, $\mathbf{P}_{blk}$, respectively, in Algorithm 2. Algorithm 2 replaces the steps 12–19 in Algorithm 1 for finding $\mathbf{q}$ and similarly for finding $\mathbf{p}$ from step 20–27. The initialization and the stopping criteria for Algorithm 2 are similar to those of Algorithm 1. Also included are the necessary steps to handle numerical errors as part of Algorithm 2 explained in Sect. 4.5. Figure 4.6 in Sect. 4.5, presents the speedup of the algorithm.

4.4.3 Bounding Existing ALS Algorithms (BALS)

In this section, we examine the algorithm for solving the Eq. (4.14) based on Theorem 4 to find the low rank factors **P**, **Q** and $\boldsymbol{\Theta}$. For the time being, assume that we need an initial, $\boldsymbol{\Theta}$, **P** and **Q** to start the algorithm. Also, we need update functions for $\boldsymbol{\Theta}$, **P**, **Q** and a stopping criteria for terminating the algorithm. The stopping criteria determines whether the constructed matrices **P** and **Q** and $\boldsymbol{\Theta}$ provide a good representation of the given matrix **R**.

input : Matrix $\mathbf{R} \in \mathbb{R}^{n \times m}$, set of indices $\mathbf{i}$, current $\mathbf{p}_x$, x, current $\mathbf{q}'_x$, r_{min}, r_{max}
output: Partial vector $\mathbf{q}_x$ of requested indices $\mathbf{i}$

// ratings of input indices $\mathbf{i}$
$\mathbf{R}_{blk} \leftarrow \mathbf{R}(:, \mathbf{i})$;
$\mathbf{M}_{blk} \leftarrow$ `ComputeRatedBinaryMatrix(`$\mathbf{R}_{blk}$`)`;
$\mathbf{P}_{blk} \leftarrow$ `Replicate(`$\mathbf{p}_x, size(\mathbf{i})$`)`;
// save $\mathbf{q}_x(\mathbf{i})$
$\mathbf{q}'_{blk} \leftarrow \mathbf{q}_x(\mathbf{i})$;
// $\mathbf{T}_{blk} \in n \times size(\mathbf{i})$ of input indices $\mathbf{i}$
$\mathbf{T}_{blk} \leftarrow \sum_{j=1, j \neq x}^{k} \mathbf{p}_j \mathbf{q}_{blk}^{\intercal}$;
// Find matrix $\mathbf{L}, \mathbf{U} \in \mathbb{R}^{n \times size(\mathbf{i})}$ as in Definition 1
$\mathbf{L} \leftarrow$ `LowerBounds(`$r_{min}, r_{max}, \mathbf{T}, \mathbf{i}, \mathbf{p}_x$`)`;
$\mathbf{U} \leftarrow$ `UpperBounds(`$r_{min}, r_{max}, \mathbf{T}, \mathbf{i}, \mathbf{p}_x$`)`;
// Find vector $\mathbf{q}_{blk}$ fixing $\mathbf{p}_x$ as in Theorem 3
$\mathbf{q}_{blk} = ([\mathbf{M}_{blk} \cdot *(\mathbf{R}_{blk} - \mathbf{T}_{blk})]^{\intercal} \mathbf{P}_{blk}) / (\|\mathbf{M}_{blk} \cdot *\mathbf{P}_{blk}\|_F^2)$;
// Find indices of $\mathbf{q}_{blk}$ that are not within bounds
$\mathbf{idxlb} \leftarrow find(\mathbf{q}_{blk} < max(\mathbf{L}))$;
$\mathbf{idxub} \leftarrow find(\mathbf{q}_{blk} > min(\mathbf{U}))$;
// case A & B numerical errors in Sect. 4.5
$idxcase1 \leftarrow find([\mathbf{q}'_{blk} \approx max(\mathbf{L})] or [\mathbf{q}'_{blk} \approx min(\mathbf{U})])$;
$idxcase2 \leftarrow find([max(\mathbf{L}) \approx min(\mathbf{U})] or [max(\mathbf{L}) > min(\mathbf{U})])$;
$idxdontchange \leftarrow idxcase1 \cup idxcase2$;
// set appropriate values of $\mathbf{q}_{blk} \notin [max(\mathbf{L}), min(\mathbf{U})]$
$\mathbf{q}_{blk}(\mathbf{idxlb} \setminus idxdontchange) \leftarrow max(\mathbf{L})(idxlb \setminus idxdontchange)$;
$\mathbf{q}_{blk}(\mathbf{idxub} \setminus idxdontchange) \leftarrow min(\mathbf{U})(idxub \setminus idxdontchange)$;
$\mathbf{q}_{blk}(idxdontchange) \leftarrow \mathbf{q}'_{blk}(idxdontchange)$;

Algorithm 2: Block BMA

input : Matrix $\mathbf{R} \in \mathbb{R}^{n \times m}$, $r_{min}, r_{max} > 1$, reduced rank k
output: Parameters $\boldsymbol{\Theta}$, Matrix $\mathbf{P} \in \mathbb{R}^{n \times k}$ and $\mathbf{Q} \in \mathbb{R}^{k \times m}$

// Rand initialization of Θ, $\mathbf{P}$, $\mathbf{Q}$.
Initialize $\boldsymbol{\Theta}, \mathbf{P}, \mathbf{Q}$ *as a random matrix* ;
$\mathbf{M} \leftarrow$ `ComputeRatedBinaryMatrix(`$\mathbf{R}$`)`;
// Compute $\mathbf{Z}$ as in Theorem 4
$\mathbf{Z} \leftarrow$ `ComputeZ(`$\mathbf{R}, \mathbf{P}, \mathbf{Q}, \boldsymbol{\Theta}$`)`;
while *stopping criteria not met* **do**
  $\boldsymbol{\theta}_i \leftarrow \underset{\boldsymbol{\theta}_i}{\operatorname{argmin}} \|\mathbf{M} \cdot *(\mathbf{R} - f(\boldsymbol{\theta}_i, \mathbf{P}, \mathbf{Q}))\|_F^2 \quad \forall 1 \le i \le l$;
  $\mathbf{P} \leftarrow \underset{\mathbf{P}}{\operatorname{argmin}} \|\mathbf{M} \cdot *(\mathbf{R} - f(\boldsymbol{\theta}_i, \mathbf{P}, \mathbf{Q}))\|_F^2$;
  $\mathbf{Q} \leftarrow \underset{\mathbf{Q}}{\operatorname{argmin}} \|\mathbf{M} \cdot *(\mathbf{R} - f(\boldsymbol{\theta}_i, \mathbf{P}, \mathbf{Q}))\|_F^2$;
  $\mathbf{Z} \leftarrow$ `ComputeZ(`$\mathbf{R}, \mathbf{P}, \mathbf{Q}, \boldsymbol{\Theta}$`)`;
  if $z_{ij} > r_{max}$ **then**
    $z_{ij} = r_{max}$
  if $z_{ij} < r_{min}$ **then**
    $z_{ij} = r_{min}$

Algorithm 3: Bounding Existing ALS Algorithm (BALS)

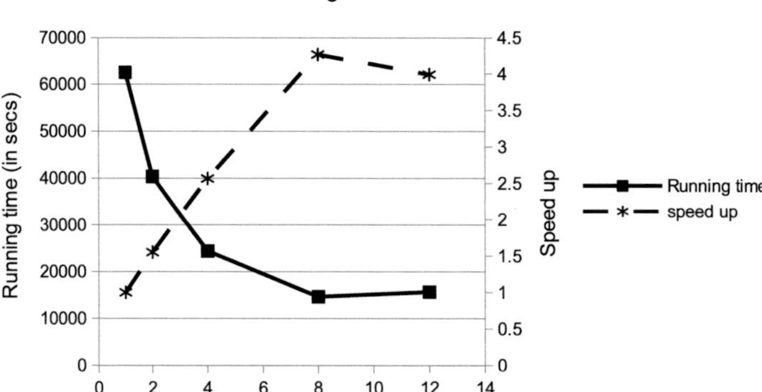

Fig. 4.6 Speedup experimentation for Algorithm 2

In the *ComputeZ* function, if the values of $\mathbf{Z}$ are outside $[r_{min}, r_{max}]$, i.e., $\mathbf{Z} > r_{max}$ and $\mathbf{Z} < r_{min}$, set the corresponding values of $\mathbf{Z} = r_{max}$ and $\mathbf{Z} = r_{min}$ respectively.

Most of the ALS-based recommender system algorithms have clearly defined blocks on $\boldsymbol{\Theta}$, $\mathbf{P}$, $\mathbf{Q}$ as discussed in Sect. 4.3.1. That is, either they are partitioned as a matrix, vector, or element blocks. Also, there is always an update order that is adhered to. For example, $\boldsymbol{\theta}_1 \rightarrow \boldsymbol{\theta}_2 \rightarrow \cdots \boldsymbol{\theta}_l \rightarrow \mathbf{p}_1 \rightarrow \mathbf{p}_2 \rightarrow \cdots \rightarrow \mathbf{p}_k \rightarrow \mathbf{q}_1 \rightarrow \mathbf{q}_2 \rightarrow \cdots \rightarrow \mathbf{q}_k$. If the algorithm meets these characteristics, we can prove that the algorithm converges to a stationary point.

Corollary 1 *If the recommender system algorithm f based on alternating least squares, satisfies the following characteristics: (1) is an iterative coordinate descent algorithm, (2) defines blocks over the optimization variables $\boldsymbol{\Theta}, \mathbf{P}, \mathbf{Q}$, and (3) has orderly optimal block updates of one block at a time and always uses the latest blocks, f converges to a stationary point of (4.14).*

Proof This is based on Theorem 1. The BALS Algorithm 3 for the formulation (4.14) satisfies the above characteristics and hence the algorithm converges to a stationary point. □

At this juncture, it is important to discuss the applicability of BMA for stochastic gradient descent type of algorithms such as SVD++, Bias-SVD, time-SVD++, and others. For brevity, we consider the SGD for the simple matrix factorization problem explained in Eq. (4.1). For a gradient descent type of algorithm, we update the current value of $p_{uk} \in \mathbf{P}$ and $q_{ki} \in \mathbf{Q}$ based on the gradient of the error $e_{ui} = r_{ui} - f(\mathbf{P}', \mathbf{Q}')$. Assuming $\lambda_p = \lambda_q = \lambda$, the update equations are $p_{uk} = p'_{uk} + (e_{ui}q'_{ki} - \lambda p'_{uk})$ and $q_{ki} = q'_{ki} + (e_{ui} * p'_{uk} - \lambda q'_{ki})$. In the case of BALS for unobserved pairs (u, i), we use $z_{ui} = f(\mathbf{P}', \mathbf{Q}')$ instead of $r_{ui} = 0$. Thus, in the case of the BALS extended gradient descent algorithms, the error $e_{ui} = 0$, for unobserved pairs (u, i). That is, the updates of p_{uk} and q_{ki} are dependent only on the observed entries and our estimates for unrated pairs (u, i) do not have any impact.

It is also imperative to understand that we are considering only the existing recommender system algorithms that minimizes the RMSE error of the observed entries against its estimation as specified in (4.2). We have not analyzed the problems that utilize different loss functions such as KL-Divergence.

In this section, we also present an example for bounding an existing algorithm, alternating least squares, with regularization in Graphchi. Graphchi [21] is an open source library that allows distributed processing of very large graph computations on commodity hardware. It breaks large graphs into smaller chunks and uses a parallel sliding window method to execute large data mining and machine learning programs in small computers. As part of the library samples, it provides implementations of many recommender system algorithms. In this section, we discuss extending the existing ALSWR implementation in Graphchi to impose box constraints using our framework.

Graphchi programs are written in the vertex-centric model and runs vertex-centric programs asynchronously (i.e., changes written to edges are immediately visible to subsequent computation), and in parallel. Any Graphchi program has three important functions: (1) *beforeIteration*, (2) *afterIteration*, and (3) *update* function. The function *beforeIteration* and *afterIteration* are executed sequentially in a single core, whereas the *update* function for all the vertices is executed in parallel across multiple cores of the machine. Such parallel updates are useful for updating independent blocks. For example, in our case, every vector $\mathbf{p}_i^\mathsf{T}$ is independent of $\mathbf{p}_j^\mathsf{T}$, for $i \neq j$.

Graphchi models the collaborative filtering problem as a bipartite graph between a user vertex and item vertex. The edges that flow between the user-item partition are ratings. That is, a weighted edge, between a user u and an item i vertex represents the user u's rating for the item i. The user vertex has only outbound edges and an item vertex has only inbound edges. The *update* function is called for all the user and item vertices. The vertex *update* solves a regularized least squares system, with neighbors' latent factors as input.

One of the major challenges for existing algorithms to enforce bounds using the BALS framework is memory. The matrix $\mathbf{Z}$ is dense and may not be accommodative in memory. That is, consider the $\mathbf{Z}$ for the book crossing dataset.[1] The dataset provides ratings of 278,858 users against 271,379 books. The size of $\mathbf{Z}$ for such a dataset would be $\mathit{number\ of\ users} * \mathit{number\ of\ items} * \mathit{size(double)} = 278{,}858 * 271{,}379 * 8$ bytes $\approx 563\,\text{GB}$. This data size is too large even for a server system. To overcome this problem, we save the $\boldsymbol{\Theta}, \mathbf{P}, \mathbf{Q}$ of previous iterations as $\boldsymbol{\Theta}', \mathbf{P}', \mathbf{Q}'$. Instead of having the entire $\mathbf{Z}$ matrix in memory, we compute the z_{ui} during the update function.

In the interest of space, we present the pseudo code for the *update* function alone. In the *beforeIteration*, function, we backup the existing variables $\boldsymbol{\Theta}, \mathbf{P}, \mathbf{Q}$ as $\boldsymbol{\Theta}', \mathbf{P}', \mathbf{Q}'$. The *afterIteration* function computes the RMSE of the validation/training set and determines the stopping criteria.

Considering Algorithm 4, it is important to observe that we use the previous iteration's $\mathbf{P}', \mathbf{Q}', \boldsymbol{\Theta}'$ only for the computation of $\mathbf{Z}$. However, for $\mathbf{P}, \mathbf{Q}$ updates, the current latest blocks are used. Also, we cannot store the matrix $\mathbf{M}$ in memory. We

[1]The details about this dataset can be found in Table 4.2.

input : Vertex v user/item, GraphContext ctx, $\mathbf{Q}, \mathbf{P}', \mathbf{Q}', \boldsymbol{\Theta}'$
output: The uth row of $\mathbf{P}$ matrix $\mathbf{p}_u^\mathsf{T} \in \mathbb{R}^k$ or the ith column $\mathbf{q}_i \in \mathbb{R}^k$
// u^{th} row of matrix Z based on Theorem 4. $\boldsymbol{\Theta}', \mathbf{P}', \mathbf{Q}'$ are the $\boldsymbol{\Theta}, \mathbf{P}, \mathbf{Q}$ from previous iteration.
// Whether the vertex is a user/item vertex is determined by the number of incoming/outgoing edges. For user vertex the number of incoming edges = 0 and for item vertex the number of outgoing edges = 0
if *vertex v is user u vertex* **then**
  // update $\mathbf{p}_u^\mathsf{T}$
  $\mathbf{z}_u^\mathsf{T} \in \mathbb{R}^m \leftarrow f(\mathbf{P}', \mathbf{Q}')$;
  // We are replacing the $\mathbf{a}_u$ in the original algorithm with $\mathbf{z}_u$
  $\mathbf{p}_u^\mathsf{T} \leftarrow (\mathbf{Q}\mathbf{Q}^\mathsf{T}) \setminus (\mathbf{z}_u^\mathsf{T} * \mathbf{Q}^\mathsf{T})^\mathsf{T}$;
else
  // update $\mathbf{q}_i$
  $\mathbf{z}_i \in \mathbb{R}^n \leftarrow f(\mathbf{P}', \mathbf{Q}')$;
  // We are replacing the $\mathbf{a}_i$ in the original algorithm with $\mathbf{z}_i$
  $\mathbf{q}_i \leftarrow (\mathbf{P}^\mathsf{T}\mathbf{P}) \setminus (\mathbf{P}^\mathsf{T} * \mathbf{z}_i)$;

Algorithm 4: update function

know that Graphchi, as part of the *Vertex* information, passes the set of incoming and outgoing edges to and from the vertex. The set of outgoing edges from the user vertex u to the item vertex v, provides information regarding the items rated by the user u. Thus, we use this information rather than maintaining $\mathbf{M}$ in memory. The performance comparison between ALSWR and ALSWR bounded on the Netflix dataset (see footnote 1) is presented in Table 4.5. Similarly, we also bounded probabilistic matrix factorization (PMF) in the Graphchi library and compared the performances of bounded ALSWR and bounded PMF algorithms using the BALS framework with its artificially truncated version on various real-world datasets (see Table 4.6).

4.4.4 Parameter Tuning

In the case of recommender systems, the missing ratings are provided as ground truth in the form of test data. The dot product of $\mathbf{P}(u, :)$ and $\mathbf{Q}(:, i)$ gives the missing rating of a (u, i) pair. In such cases, the accuracy of the algorithm is determined by the root mean square error (RMSE) of the predicted ratings against the ground truth. How well the algorithm converges for a given k is unimportant.

This section discusses ways to improve the RMSE of the predictions against the missing ratings by tuning the parameters of the BMA algorithm and BALS framework.

4.4.4.1 Initialization

The BMA algorithm can converge to different points depending on the initialization. In Algorithm 1, it was shown how to use random initialization so that $\mathbf{PQ} \in [r_{\min}, r_{\max}]$. In general, this method should provide good results.

However, in the case of recommender systems, this initialization can be tuned, which can give even better results. According to Koren [17], one good baseline estimate for a missing rating (u, i) is $\mu + g_u + h_i$, where μ is the average of the known ratings, and g_u and h_i are the bias of user u and item i, respectively. We initialized $\mathbf{P}$ and $\mathbf{Q}$ in the following way

$$\mathbf{P} = \begin{pmatrix} \frac{\mu}{k-2} & \cdots & \frac{\mu}{k-2} & g_1 & 1 \\ \vdots & & \vdots & \vdots & \vdots \\ \frac{\mu}{k-2} & \cdots & \frac{\mu}{k-2} & g_n & 1 \end{pmatrix} \quad \text{and} \quad \mathbf{Q} = \begin{pmatrix} 1 & 1 & \cdots & 1 \\ \vdots & \vdots & & \vdots \\ 1 & 1 & \cdots & 1 \\ h_1 & h_2 & \cdots & h_m \end{pmatrix},$$

such that $\mathbf{PQ}(u, i) = \mu + g_u + h_i$. That is, let the first $k-2$ columns of $\mathbf{P}$ be $\frac{\mu}{k-2}$, $\mathbf{P}(:, k-1) = \mathbf{g}$ and $\mathbf{P}(:, k) = 1$. Let all the $k-1$ rows of $\mathbf{Q}$ be 1's and $\mathbf{Q}(k, :) = \mathbf{h}^\mathsf{T}$. We call this a baseline initialization.

4.4.4.2 Reduced Rank k

In the case of regular low rank approximation with all known elements, the higher the k, the closer the low rank approximation is to the input matrix [15]. However, in the case of predicting with the low rank factors, a good k depends on the nature of the dataset. Even though, for a higher k, the low rank approximation is closer to the known rating of the input $\mathbf{R}$, the RMSE on the test data may be poor. In Table 4.3, we can observe the behavior of the RMSE on the test data against k. In most cases, a good k is determined by trial and error for the prediction problem.

4.4.4.3 Stopping Criterion $\mathfrak{C}$

The stopping criterion defines the goodness of the low rank approximation for the given matrix and the task for which the low rank factors are used. The two common stopping criteria are—(1) For a given rank k, the product of the low rank factors $\mathbf{PQ}$ should be close to the known ratings of the matrix and (2) The low rank factors $\mathbf{P}, \mathbf{Q}$ should perform the prediction task on a smaller validation set which has the same distribution as the test set. The former is common when all the elements of $\mathbf{R}$ are known. We discuss only the latter, which is important for recommender systems.

The stopping criterion $\mathfrak{C}$ for the recommender system is the increase of $\sqrt{\frac{\|\mathbf{M}\cdot*(\mathbf{V}-\mathbf{PQ})\|_F^2}{numRatings\ in\ \mathbf{V}}}$, for some validation matrix $\mathbf{V}$, which has the same distribution as the test matrix between successive iterations. Here, $\mathbf{M}$ is for the validation matrix $\mathbf{V}$. This stopping criterion has diminishing effect as the number of iterations increases. Hence, we also check whether $\sqrt{\frac{\|\mathbf{M}\cdot*(\mathbf{V}-\mathbf{PQ})\|_F^2}{numRatings\ in\ \mathbf{V}}}$ did not change in successive iterations at a given floating point precision, e.g., 1e-5.

It is trivial to show that, for the above stopping criterion $\mathfrak{C}$, Algorithm 1 terminates for any input matrix $\mathbf{R}$. At the end of an iteration, we terminate if the RMSE on the validation set has either increased or marginally decreased.

4.5 Experimentation

Experimentation was conducted in various systems with memory as low as 16 GB. One of the major challenges during experimentation is numerical errors. The numerical errors could result in $\mathbf{T}+\mathbf{p}_x\mathbf{q}_x^\mathsf{T} \notin [r_{min}, r_{max}]$. The two fundamental questions to solve the numerical errors are: (1) How to identify the occurrence of a numerical error? and (2) What is the best possible value to choose in the case of a numerical error?

We start by addressing the former question of potential numerical errors that arise in the BMA Algorithm 1. It is important to understand that if we are well within bounds, i.e., if $max(\mathbf{l}) < q_{xi} < min(\mathbf{u})$, we are not essentially impacted by the numerical errors. It is critical only when q_{xi} is out of the bounds, that is, $q_{xi} < max(\mathbf{l})$ or $q_{xi} > min(\mathbf{u})$ *and* approximately closer to the boundary discussed as in (*Case A* and *Case B*). For discussion, let us assume we are improving the old value of q'_{xi} to q_{xi} such that we minimize the error $\|\mathbf{M}\cdot*(\mathbf{R}-\mathbf{T}-\mathbf{p}_x\mathbf{q}_x^\mathsf{T})\|_F^2$.

Case A: $q'_{xi} \approx max(\mathbf{l})$ or $q'_{xi} \approx min(\mathbf{u})$:
This is equivalent to saying q'_{xi} is already optimal for the given $\mathbf{p}_x$ and $\mathbf{T}$ and there is no further improvement possible. Under this scenario, if $q'_{xi} \approx q_{xi}$ it is better to retain q'_{xi} irrespective of the new q_{xi} found.

Case B: $max(\mathbf{l}) \approx min(\mathbf{u})$ or $max(\mathbf{l}) > min(\mathbf{u})$:
According to Theorem 2, we know that $max(\mathbf{l}) < min(\mathbf{u})$. Hence, if $max(\mathbf{l}) > min(\mathbf{u})$, it is only the result of numerical errors.

In all the above cases during numerical errors, we are better off retaining the old value q'_{xi} against the new value q_{xi}. This covers Algorithm 2—Block BMA for consideration of numerical errors.

We experimented with Algorithm 2 among varied bounds using very large matrix sizes taken from the real-world datasets. The datasets used for our experiments included the Movielens 10 million [26], Jester [7], Book crossing [32] and Online dating dataset [2]. The characteristics of the datasets are presented in Table 4.2.

Table 4.2 Datasets for experimentation

Dataset	Rows	Columns	Ratings (millions)	Density	Ratings range
Jester	73,421	100	4.1	0.5584	[−10, 10]
Movielens	71,567	10,681	10	0.0131	[1, 5]
Dating	135,359	168,791	17.3	0.0007	[1, 10]
Book crossing	278,858	271,379	1.1	0.00001	[1, 10]
Netflix	17,770	480,189	100.4	0.01	[1, 5]

We have chosen root mean square error (RMSE)—a defacto metric for recommender systems. The RMSE is compared for BMA with baseline initialization (BMA—Baseline) and BMA with random initialization (BMA—Random) against the other algorithms on all the datasets. The algorithms used for comparison are ALSWR (alternating least squares with regularization) [31], SGD [6], SVD++ [17], and Bias-SVD [17] and its implementation in the Graphlab (http://graphlab.org/) [23] software package. We implemented our algorithm in MATLAB and used the parallel computing toolbox for parallelizing across multiple cores.

For parameter tuning, we varied the number of reduced rank k and tried different initial matrices for our algorithm to compare against all other algorithms mentioned above. For every k, every dataset was randomly partitioned into 85 % training, 5 % validation, and 10 % test data. We ran all algorithms on these partitions and computed their RMSE scores. We repeated each experiment 5 times and reported their RMSE scores in Table 4.3, where each resulting value is the average of the RMSE scores on a randomly chosen test set for 5 runs. Table 4.3 summarizes the RMSE comparison of all the algorithms.

Table 4.3 RMSE comparison of algorithms on real-world datasets

Dataset	k	BMA baseline	BMA random	ALSWR	SVD++	SGD	Bias-SVD
Jester	10	4.3320	4.6289	5.6423	5.5371	5.7170	5.8261
Jester	20	4.3664	4.7339	5.6579	5.5466	5.6752	5.7862
Jester	50	4.5046	4.7180	5.6713	5.5437	5.6689	5.7956
Movielens10M	10	0.8531	0.8974	1.5166	1.4248	1.2386	1.2329
Movielens10M	20	0.8526	0.8931	1.5158	1.4196	1.2371	1.2317
Movielens10M	50	0.8553	0.8932	1.5162	1.4204	1.2381	1.2324
Dating	10	1.9309	2.1625	3.8581	4.1902	3.9082	3.9052
Dating	20	1.9337	2.1617	3.8643	4.1868	3.9144	3.9115
Dating	50	1.9434	2.1642	3.8606	4.1764	3.9123	3.9096
Book crossing	10	1.9355	2.8137	4.7131	4.7315	5.1772	3.9466
Book crossing	20	1.9315	2.4652	4.7212	4.6762	5.1719	3.9645
Book crossing	50	1.9405	2.1269	4.7168	4.6918	5.1785	3.9492

Algorithm 1 consistently outperformed existing state-of-the-art algorithms. One of the main reasons for the consistent performance is the absence of hyperparameters. In the case of machine learning algorithms, there are many parameters that need to be tuned for performance. Even though the algorithms perform the best when provided with the right parameters, identifying these parameters is a formidable challenge, usually by trial and error methods. For example, in Table 4.3, we can observe that the Bias-SVD, an algorithm without hyperparameters, performed better than its extension SVD++ with default parameters in many cases. The BMA algorithm without hyperparameters performed well on real-world datasets, albeit a BMA with hyperparameters and the right parametric values would have performed even better.

Recently, there has been a surge in interest to understand the temporal impact on the ratings. Time-svd++ [18] is one such algorithm that leverages the time of rating to improve prediction accuracy. Also, the most celebrated dataset in the recommender system community is the Netflix dataset, since the prize money is attractive and it represents the first massive dataset for recommender systems that was publicly made available. The Netflix dataset consists of 17,770 users who rated 480,189 movies in a scale of [1, 5]. There was a total of 100,480,507 ratings in the training set and 1,408,342 ratings in the validation set. All the algorithms listed above were invented to address the Netflix challenge. Even though the book crossing dataset [32] is bigger than the Netflix, we felt our study is not complete without experimenting on Netflix and comparing against time-SVD++. However, the major challenge is that the Netflix dataset has been withdrawn from the internet and its test data is no longer available. Hence, we extracted a small sample of 5 % from the training data as a validation set and tested the algorithm against the validation set that was supplied as part of the training package. We performed this experiment and the results are presented in Table 4.4. For a better comparison, we also present the original Netflix test scores for SVD++ and time-SVD++ algorithms from [18]. These are labeled *SVD++-Test* and *time-SVD++-Test*, respectively. Our BMA algorithm outperformed all the algorithms on the Netflix dataset when tested on the validation set supplied as part of the Netflix training package.

Table 4.4 RMSE comparison of BMA with other algorithms on Netflix

Algorithm	$k = 10$	$k = 20$	$k = 50$	$k = 100$
BMA baseline	0.9521	0.9533	0.9405	0.9287
BMA random	0.9883	0.9569	0.9405	0.8777
ALSWR	1.5663	1.5663	1.5664	1.5663
SVD++	1.6319	1.5453	1.5235	1.5135
SGD	1.2997	1.2997	1.2997	1.2997
Bias-SVD	1.3920	1.3882	1.3662	1.3354
Time-svd++	1.1800	1.1829	1.1884	1.1868
SVD++-test	0.9131	0.9032	0.8952	0.8924
Time-SVD++-test	0.8971	0.8891	0.8824	0.8805

Additionally, we conducted an experiment to study the speedup of the algorithm on the Netflix dataset. This is a simple speedup experiment conducted with MATLAB's Parallel Computing Toolbox on a dual socket Intel E7 system with 6 cores on each socket. We collected the running time of the Algorithm 2 to compute the low rank factors **P** and **Q** with $k = 50$, using $1,\ 2,\ 4,\ 8$, and 12 parallel processes. MATLAB's Parallel Computing Toolbox permits starting at most 12 MATLAB workers for a local cluster. Hence, we conducted the experiment up to a pool size of 12. Figure 4.6 shows the speedup of Algorithm 2. We observe from the graph that, up to pool size 8, the running time decreases with increasing pool size. However, the overhead costs such as communication and startup costs for running 12 parallel tasks surpasses the advantages of parallel execution. This simple speedup experiment shows promising reductions in running time of the algorithm. A sophisticated implementation of the algorithm with low-level parallel programming interfaces such as MPI, will result in better speedups.

In this section, we also present the results of bounding existing ALS-type algorithms as explained in Sects. 4.3.3 and 4.4.3. The performance comparison between ALSWR and ALSWR bounded on the Netflix dataset is presented in Table 4.5. Similarly, we also bounded Probabilistic Matrix Factorization (PMF) in the Graphchi library. We then compared the performances of both ALSWR and PMF algorithms on various real-world datasets, which are presented in Table 4.6.

Table 4.5 RMSE comparison of ALSWR on Netflix

Algorithm	$k = 10$	$k = 20$	$k = 50$
ALSWR	0.8078	0.755	0.6322
ALSWR bounded	0.8035	0.7369	0.6156

Table 4.6 RMSE comparison using BALS framework on real-world datasets

Dataset	k	ALSWR bounded	PMF bounded	ALSWR	PMF
Jester	10	4.4406	4.2011	4.4875	4.2949
Jester	20	4.8856	4.3018	5.0288	4.4608
Jester	50	5.6177	4.6893	6.1906	4.7383
ML-10M	10	0.8869	0.8611	0.9048	0.8632
ML-10M	20	0.9324	0.8752	0.9759	0.8891
ML-10M	50	1.0049	0.8856	1.1216	0.9052
Dating	10	2.321	1.9503	2.3206	1.9556
Dating	20	2.3493	1.9652	2.4458	1.9788
Dating	50	2.7396	2.0647	2.7406	2.0752
Book crossing	10	4.6937	5.4676	4.7805	5.4901
Book crossing	20	4.7977	5.3977	4.8889	5.4862
Book crossing	50	5.0102	5.2281	5.0018	5.4707

4.6 Conclusion

In this chapter, we presented a new matrix factorization for recommender systems called Bounded Matrix Low Rank Approximation (BMA), which imposes a lower and an upper bound on every estimated missing element of the input matrix. Also, we presented substantial experimental results on real-world datasets illustrating that our proposed method outperformed the state-of-the-art algorithms for recommender systems.

In future work, we plan to extend BMA to tensors, i.e., multi-way arrays. Also, similar to time-SVD++, we will use time, neighborhood information, and implicit ratings during the factorization. A major challenge of the BMA algorithm is that it loses sparsity during the product of low rank factors **PQ**. This limits the applicability of BMA to other datasets such as text corpora and graphs where sparsity is important. Thus, we plan to extend BMA for sparse bounded input matrices as well. During our experimentation, we observed linear scale-up for Algorithm 2 in MATLAB. However, the other algorithms from Graphlab are implemented in C/C++ and take less clock time. A C/C++ implementation of Algorithm 2 would be an important step in order to compare the running time performance against the other state-of-the-art algorithms. Also, we will experiment with BMA on other types of datasets that go beyond those designed for recommender systems.

Acknowledgments This work was supported in part by the NSF Grant CCF-1348152, the Defense Advanced Research Projects Agency (DARPA) XDATA program grant FA8750-12-2-0309, Research Foundation Flanders (FWO-Vlaanderen), the Flemish Government (Methusalem Fund, METH1), the Belgian Federal Government (Interuniversity Attraction Poles—IAP VII), the ERC grant 320378 (SNLSID), and the ERC grant 258581 (SLRA). Mariya Ishteva is an FWO Pegasus Marie Curie Fellow. Any opinions, findings, and conclusions or recommendations expressed in this material are those of the authors and do not necessarily reflect the views of the NSF or the DARPA.

References

1. D.P. Bertsekas, *Nonlinear Programming* (Athena Scientific, Belmont, 1999)
2. L. Brozovsky, V. Petricek, Recommender system for online dating service, in *Proceedings of Conference Znalosti 2007*, Ostrava, VSB (2007)
3. A. Cichocki, A.-H. Phan, Fast local algorithms for large scale nonnegative matrix and tensor factorizations. IEICE Trans. Fundam. Electron. Commun. Comput. Sci. **E92–A**, 708–721 (2009)
4. A. Cichocki, R. Zdunek, S. Amari, Hierarchical ALS algorithms for nonnegative matrix and 3d tensor factorization. Lect. Notes Comput. Sci. **4666**, 169–176 (2007)
5. S. Deerwester, S.T. Dumais, G.W. Furnas, T.K. Landauer, R. Harshman, Indexing by latent semantic analysis. J. Am. Soc. Inf. Sci. **41**, 391–407 (1990)
6. S. Funk, Stochastic gradient descent. (2006) http://sifter.org/simon/journal/20061211.html [Online; accessed 6-June-2012]
7. K. Goldberg, Jester collaborative filtering dataset. (2003) http://goldberg.berkeley.edu/jester-data/ [Online; accessed 6-June-2012]

8. G.H. Golub, C.F. Van Loan, *Matrix Computations*, 3rd edn. (The Johns Hopkins University Press, Baltimore, 1996)
9. L. Grippo, M. Sciandrone, On the convergence of the block nonlinear Gauss-Seidel method under convex constraints. Oper. Res. Lett. **26**(3), 127–136 (2000)
10. N.-D. Ho, P.V. Dooren, V.D. Blondel, Descent methods for nonnegative matrix factorization. (2008) CoRR, abs/0801.3199
11. R. Kannan, M. Ishteva, H. Park, Bounded matrix low rank approximation, in *Proceedings of the 12th IEEE International Conference on Data Mining(ICDM-2012)* (2012), pp. 319–328
12. R. Kannan, M. Ishteva, H. Park, Bounded matrix factorization for recommender system. Knowl. Inf. Syst. **39**(3), 491–511 (2014)
13. H. Kim, H. Park, Sparse non-negative matrix factorizations via alternating non-negativity-constrained least squares for microarray data analysis. Bioinformatics **23**(12), 1495–1502 (2007)
14. H. Kim, H. Park, Nonnegative matrix factorization based on alternating nonnegativity constrained least squares and active set method. SIAM J. Matrix Anal. Appl. **30**(2), 713–730 (2008)
15. J. Kim, Y. He, H. Park, Algorithms for nonnegative matrix and tensor factorizations: a unified view based on block coordinate descent framework. J. Glob. Optim. 1–35 (2013)
16. J. Kim, H. Park, Fast nonnegative matrix factorization: an active-set-like method and comparisons. SIAM J. Sci. Comput. **33**(6), 3261–3281 (2011)
17. Y. Koren, Factorization meets the neighborhood: a multifaceted collaborative filtering model, in *Proceeding of the 14th ACM SIGKDD International Conference on Knowledge Discovery and Data Mining—KDD'08* (2008), pp. 426–434
18. Y. Koren, Collaborative filtering with temporal dynamics, in *Proceedings of the 15th ACM SIGKDD International Conference on Knowledge Discovery and Data Mining—KDD'09* (2009), pp. 447
19. Y. Koren, R. Bell, C. Volinsky, Matrix factorization techniques for recommender systems. Computer **42**(8), 30–37 (2009)
20. D. Kuang, H. Park, C.H.Q. Ding, Symmetric nonnegative matrix factorization for graph clustering, in *Proceedings of SIAM International Conference on Data Mining—SDM'12* (2012), pp. 106–117
21. A. Kyrola, G. Blelloch, C. Guestrin, Graphchi: large-scale graph computation on just a PC, in *Proceedings of the 10th USENIX Conference on Operating Systems Design and Implementation, OSDI'12*, (USENIX Association, Berkeley, 2012) pp. 31–46
22. C.J. Lin, Projected gradient methods for nonnegative matrix factorization. Neural Comput. **19**(10), 2756–2779 (2007)
23. Y. Low, J. Gonzalez, A. Kyrola, D. Bickson, C. Guestrin, J.M. Hellerstein, Graphlab: a new parallel framework for machine learning, in *Conference on Uncertainty in Artificial Intelligence (UAI)* (2010)
24. L.W. Mackey, D. Weiss, M.I. Jordan, Mixed membership matrix factorization, in *Proceedings of the 27th International Conference on Machine Learning (ICML-10)* (2010), pp. 711–718
25. I. Markovsky, Algorithms and literate programs for weighted low-rank approximation with missing data, in *Approximation Algorithms for Complex Systems*, ed. by J. Levesley, A. Iske, E. Georgoulis (Springer, Berlin, 2011), pp. 255–273. Chap. 12
26. Movielens dataset. (1999) http://movielens.umn.edu [Online; accessed 6-June-2012]
27. A. Paterek, Improving regularized singular value decomposition for collaborative filtering, in *Proceedings of 13th ACM International Conference on Knowledge Discovery and Data Mining—KDD'07* (2007), pp. 39–42
28. R. Salakhutdinov, A. Mnih, Bayesian probabilistic matrix factorization using Markov chain Monte Carlo, in *ICML* (2008), pp. 880–887
29. L. Xiong, X. Chen, T.-K. Huang, J.G. Schneider, J.G. Carbonell, Temporal collaborative filtering with Bayesian probabilistic tensor factorization, in *Proceedings of the SIAM International Conference on Data Mining-SDM'10* (2010), pp. 211–222

30. H.-F. Yu, C.-J. Hsieh, S. Si, I.S. Dhillon, Scalable coordinate descent approaches to parallel matrix factorization for recommender systems, in *Proceedings of the IEEE International Conference on Data Mining-ICDM'12* (2012), pp. 765–774
31. Y. Zhou, D. Wilkinson, R. Schreiber, R. Pan, Large-scale parallel collaborative filtering for the Netflix prize. Algorithm. Asp. Inf. Manag. **5034**, 337–348 (2008)
32. C.-N. Ziegler, S.M. McNee, J.A. Konstan, G. Lausen, Improving recommendation lists through topic diversification, in *Proceedings of the 14th International Conference on World Wide Web-WWW'05* (2005), pp. 22–32

Chapter 5
A Modified NMF-Based Filter Bank Approach for Enhancement of Speech Data in Nonstationary Noise

Farook Sattar and Feng Jin

Abstract This article addresses the problem of single-channel speech enhancement in the presence of nonstationary noise. A novel-modified NMF-based filter bank approach is proposed for speech enhancement. The method consists of filter bank analysis of the noisy input signal followed by extraction of speech signal based on a modified NMF (MNMF) by learning a speaker-independent speech dictionary using a precomputed noise dictionary. The proposed method works well with the real-world nonstationary noise independent to the speaker. The method is evaluated using a speech database consists of different speakers showing promising enhancement performance by reducing the nonstationary noise together with improving PESQ (perceptual evaluation of speech quality) compared to other competitive state-of-the-art methods.

Keywords Single-channel speech enhancement · Modified NMF · Filter bank · Nonstationary noise · Noizeus speech corpus

5.1 Introduction

The speech enhancement in real-world nonstationary noise is a challenging problem. Traditional speech enhancement algorithms [1] based on spectral subtraction, subspace, statistical model, and Wiener filtering have limitations to work properly in highly nonstationary noise environment. For instance, the conventional algorithms have the following two restrictions. First, the noise suppression criterion is usually based on some form of segmental SNR, which is not so reliable in nonstationary

F. Sattar (✉)
Department of Electrical and Computer Engineering, University of Waterloo,
Waterloo, ON, Canada
e-mail: fsattar@uwaterloo.ca

F. Jin
Department of Electrical and Computer Engineering, Ryerson University,
Toronto, ON, Canada

G.R. Naik (ed.), *Non-negative Matrix Factorization Techniques*,
Signals and Communication Technology, DOI 10.1007/978-3-662-48331-2_5

noise condition [2]. Second, it happens that usually dominating low-frequency nonstationary noise components cannot be effectively removed using the SNR-based suppression rule, or even incorporating signal presence uncertainty [3], which is based on the minimum mean square estimate (MMSE) of the clean speech and a decision-directed a priori SNR from which a multiplicative gain for the subtraction algorithm is determined. However, the determination of the a priori SNR is difficult since very precise estimates of noise are not available. Also, the residual noise arises due to the fact that the human auditory system is highly nonlinear. Therefore, analytic methods based on mathematical criteria for estimation of a priori SNR are not able to eliminate musical noise altogether. In [4], the method calculates a noise masking threshold and adjust the subtraction parameters depending on its value in each subband. The method in [5] utilizes the noise masking based on the definition of an audible noise spectrum and its subsequent suppression using optimal nonlinear filtering of the short-time spectral amplitude (STSA) envelop.

The nonnegative matrix factorization (NMF) technique introduced by Lee and Seung [6] is a subspace method which finds a linear data representation with nonnegativity constraints. It attempts matrix factorization in a way that the factors have nonnegative elements by performing a simple multiplicative updating. NMF is able to produce useful representations of real-world data and can, therefore, be applied here to solve the problem of speech enhancement in real-world noise environment. Based on the observed noisy speech data, the NMF decomposes the data matrix into two basis matrices. This results in reduced representation of the original data where each feature is a linear combination of the original attribute set [7]. The NMF has low computational complexity and unlike time–frequency techniques, it is able to deal with both dense and sparse datasets [8, 9]. Some relevant applications of NMF include single-channel source separation for drum transcription [10], and blind recovery of single-channel mixed cardiac and respiratory sounds [11].

In [12], a regularized NMF-based oracle speech enhancement method is proposed using spectral components modeled by Gaussian mixtures with clean speech and noise data. Using the estimated GMM parameters by oracle training and log-likelihood cost function, the magnitude spectrum of the clean speech is estimated based on a regularized NMF algorithm with the basis matrices of both clean and noise. Finally, the phase of the noisy speech is combined with the estimated magnitude spectrum of the clean speech and the enhanced speech signal is reconstructed in the time domain. In [13], a segmental NMF (SNMF)-based speech enhancement scheme is proposed by considering the spectro-temporal processing of the magnitude spectrogram. The spectral processing consists of dividing the magnitude spectrogram of the speech signal into low- and high-frequency bands followed by applying NMF to each sub-matrices referring to low- and high-frequency bands. The temporal processing, which is capturing temporal information, is based on patch processing by segmenting the magnitude spectrogram into a number of frames and generating patches consist of neighboring frames and subsequently applying the NMF with the patches. Both the spectral- and temporal-domain SNMF algorithms are finally combined to enhance the noisy speech signals. In [14], a two-stage speech enhancement method is presented by integrating the statistical models and NMF algorithm with both speech and noise bases update using the speech presence probability (SPP). In stage 1, the noisy input signal is preprocessed by statistical model-based enhancement (SE) and the speech and noise bases are updated for the second stage using SPP. The stage 2 contains an NMF-based enhancement module where the minimum mean square error-log spectral amplitude (MMSE-LSA) estimator is adopted for which the signal-to-noise ratio (SNR)-related parameters are estimated by the NMF module. In [15], deep neural network (DNN)-based NMF method is proposed for speech enhancement by separating the target speech source from the mixed source. The encoding vectors of the NMF are estimated by DNN which can then reconstruct the desired source data vectors.

In this article, we propose a method that is able to enhance a speech signal in the presence of additive nonstationary noise through a combination of filter bank and MNMF for reconstructing the parts of speech signal that are obscured by the interference. By precomputing the nonstationary noise dictionary, the speaker-independent speech dictionary is learned by our modified NMF scheme from the outputs of a uniform filter bank. The proposed method is thus able to enhance the speech signal independent to the speaker and outperforms the competing relevant methods based on generalized spectral subtraction (GSS) [16] and a posteriori SNR-based t–f (time–frequency) approach [17].

5.2 Proposed Speech Enhancement Method

In the following, we propose a new method for speech enhancement based on filter bank analysis and the modified NMF.

5.2.1 Filter Bank

The noisy speech signal is first bandpass filtered into M subbands by a uniform filter bank. Basically, the purpose of a filter bank is to split the frequency band of a signal into several subbands and to process the subband signals separately. Since the frequency functions of the subbands are almost nonoverlapping, the outputs will be linear independent. The filtering process through a filter bank can be regarded as a transform operation. With a transform matrix $\mathbf{T}$ of size $(N_f \times L)$, one can transform a N_f-sample input signal into L outputs, $\phi_m(n)$, with $N_f \geq L$. A block scheme of the transform operation is shown in Fig. 5.1. The number of numerical operations can be made proportional to N_f if the transform T is well chosen using a decimation–interpolation procedure [18]. Examples of transforms which can be related to a filter bank are DFT, DCT, and WT [19, 20].

When the filter bank has filters that are twice as long as the number of subbands, the transform of a signal can be considered as to be a lapped.

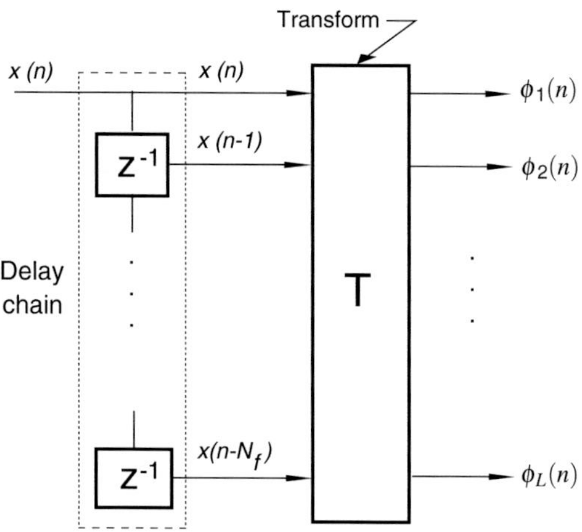

Fig. 5.1 A scheme of the transform operation

5.2.1.1 The Uniform Filter Bank

The impulse responses of a uniform filter bank with L subfilters are denoted $h_m(n)$, and the corresponding transfer functions are denoted $H_m(f)$. The impulse responses are given by

$$h_m(n) = \frac{1}{L} h_0(n) \exp\{j2\pi(m-1)n/L\}, \qquad m = 1, \ldots, L, \tag{5.1}$$

where $h_0(n)$ is the impulse response of a prototype low-pass filter with a frequency response of $H_0(f)$. The subfilters are constructed by the following relationship:

$$\left| \sum_{m=1}^{L} H_m(f) \right| = 1. \tag{5.2}$$

One prototype filter which fulfills this demand is

$$\begin{cases} \frac{1}{L} H_0(f) = \begin{cases} 1 - L\,|f| & |f| < 1/L \\ 0 & \text{else} \end{cases} \\ \\ \frac{1}{L} h_0(n) = \frac{1}{L} \left(\frac{\sin \pi n/L}{\pi n/L} \right)^2 . \end{cases} \tag{5.3}$$

The prototype filter in Eq. (5.3) is one filter, which fulfills Eq. (5.2). It will be approximated by an FIR filter of length N_f. From an implementation point of view, it is advantageous to relate the length N_f and the number L of subfilters according to $N_f = lL$, where l is an integer. An approximation of $|H_0(f)|$ in Eq. (5.3) with $l = 2$ is shown in Fig. 5.2a (dashed line), and in Fig. 5.2b, the first three bandpass subfilters

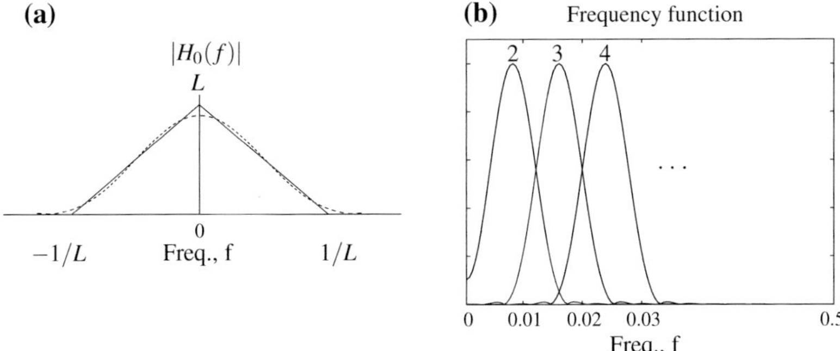

Fig. 5.2 **a** The magnitude function of the prototype low-pass filter $h_0(n)$; **b** Filter responses of the uniform filter bank

are depicted when the length of the approximation of $h_0(n)$ is $N_f = 128$ and $L = 64$. This approximation will only be slightly improved if $l > 2$. Thus, for a given L, there will be only a minor improvement of the estimate by choosing $l > 2$, i.e., by making $N_f > 2L$. The effect on the approximated impulse response is that its tails will be longer. It will, on the other hand, increase the number of numerical operations, which is proportional to N_f [20]. The quality of the estimated signal will, however, be improved by increasing L, i.e., by increasing the frequency resolution. The degree of the improvement depends on the signal to be processed.

5.2.2 Modified NMF

A modified NMF-based framework is introduced here by different analysis and synthesis dictionaries (basis functions) in the NMF scheme. The idea of nonnegative matrix factorization (NMF) [6] to decompose the data matrix, $\mathbf{V}$, as

$$\mathbf{V} \approx \mathbf{D}_v \mathbf{H}_v \tag{5.4}$$

where $\mathbf{D}_v$ and $\mathbf{H}_v$ refer to the nonnegative matrices which we call the dictionary (basis functions) and the code (coefficients) of the nonstationary noise, respectively. There exists different algorithms for computing NMF factorization [21–24]. Here, we use the method proposed in [22] due to its simple formulation and easy implementation. This NMF algorithm begins with randomly initialized matrices, $\mathbf{D}_v$ and $\mathbf{H}_v$, and alternates the following updates until convergence.

$$\mathbf{H}_v \leftarrow \mathbf{H}_v \bullet \left((\overline{\mathbf{D}}_v^T \mathbf{V})./(\overline{\mathbf{D}}_v^T \overline{\mathbf{D}}_v \mathbf{H}_v + \lambda) \right) \tag{5.5a}$$

$$\mathbf{D}_v \leftarrow \overline{\mathbf{D}_v} \bullet \left((\mathbf{V}\mathbf{H}_v^T + \overline{\mathbf{D}_v} \bullet (\mathbf{1}(\overline{\mathbf{D}_v}\mathbf{H}v\mathbf{H}_v^T \bullet \overline{\mathbf{D}_v}))). / (\overline{\mathbf{D}_v}\mathbf{H}_v\mathbf{H}_v^T + \overline{\mathbf{D}_v} \bullet (\mathbf{1}\mathbf{V}\mathbf{H}_v^T \bullet \overline{\mathbf{D}_v}))) \right) \tag{5.5b}$$

where $\overline{\mathbf{D}_v}$ is the columnwise normalized dictionary matrix, $\mathbf{1}$ is a square matrix with all elements equal to 1, '•' refers to pointwise multiplication, './' denotes the pointwise division, λ parameter represents the degree of sparsity, and T denotes the transportation.

The sparse coding framework for the noisy input signal is assumed as

$$\mathbf{X} = \mathbf{X}_s + \mathbf{X}_v \approx \underbrace{[\mathbf{D}_s \ \ \mathbf{D}_v]}_{\mathbf{D}} \underbrace{\begin{bmatrix} \mathbf{H}_s \\ \mathbf{H}_v \end{bmatrix}}_{\mathbf{H}} \tag{5.6}$$

where the subscripts, s and v, denote the speech and the noise. Here, our idea is to precompute the source dictionary of the nonstationary noise $\mathbf{D}_v$ and then learn the dictionary of the speech directly from the noisy data. So modification of the NMF becomes so that $\mathbf{D}_s$, $\mathbf{H}_s$, and $\mathbf{H}_v$ are updated as follows:

$$\mathbf{H}_s \leftarrow \mathbf{H}_s \bullet \left((\overline{\mathbf{D}}_s^T \mathbf{X})./(\overline{\mathbf{D}}_s^T \overline{\mathbf{D}}\mathbf{H} + \lambda_s) \right) \tag{5.7a}$$

$$\mathbf{D}_s \leftarrow \overline{\mathbf{D}_s} \bullet \left((\mathbf{X}\mathbf{H}_s^T + \overline{\mathbf{D}_s} \bullet (\mathbf{1}(\overline{\mathbf{D}}\mathbf{H}\mathbf{H}_s^T \bullet \overline{\mathbf{D}_s}))). / (\overline{\mathbf{D}}\mathbf{H}\mathbf{H}_s^T + \overline{\mathbf{D}_s} \bullet (\mathbf{1}\mathbf{X}\mathbf{H}_s^T \bullet \overline{\mathbf{D}_s})) \right) \tag{5.7b}$$

$$\mathbf{H}_v \leftarrow \mathbf{H}_v \bullet \left((\overline{\mathbf{D}}_v^T \mathbf{X})./(\overline{\mathbf{D}}_v^T \overline{\mathbf{D}}\mathbf{H} + \lambda_v) \right) \tag{5.7c}$$

where λ_s and λ_v are sparsity parameters for speech and noise, respectively.

We first compute the NMF of a noise-only sequence using Eqs. (5.5a) and (5.5b). Then we discard the code matrix and use the noise dictionary matrix to compute the NMF using Eqs. (5.7a), (5.7b), and (5.7c). Finally, the reconstructed enhanced speech signal $\mathbf{x}_s$ is obtained as

$$\mathbf{x}_s = \sum (\mathbf{D}_s \mathbf{H}_s)^T \tag{5.8}$$

Algorithm 1: MNMF (precompute the noise dictionary)

Require: $\mathbf{V} \in \mathbb{R}^{n \times m}$ {dataset}
Require: λ {regularization parameter}
Require: $\mathbf{H}_{vo} \in \mathbb{R}^{n \times k}$ and $\mathbf{D}_{vo} \in \mathbb{R}^{k \times m}$ {initial matrices $\mathbf{H}_v$ and $\mathbf{D}_v$}
1. Normalize $\mathbf{V}$
2. **while** stopping criterion not satisfied **do**
3. update matrix $\mathbf{H}_v$ according to (5.5a)
4. update matrix $\mathbf{D}_v$ according to (5.5b)
5. **end while**
Output: $\mathbf{D}_v$

Algorithm 2: MNMF (learning speech dictionary using the precomputed noise dictionary)

Require: $\mathbf{X} \in \mathbb{R}^{n \times m}$ {dataset}
Require: $\mathbf{D}_v \in \mathbb{R}^{n \times m}$ {dataset}
Require: λ_s {regularization parameter for speech}
Require: λ_v {regularization parameter for noise}
Require: $\mathbf{H}_{so} \in \mathbb{R}^{n \times k}$ and $\mathbf{D}_{so} \in \mathbb{R}^{k \times m}$ {initial matrices $\mathbf{H}_s$ and $\mathbf{D}_s$}
1. Normalize $\mathbf{X}$
2. **while** stopping criterion not satisfied **do**
3. form the $\mathbf{D}$ martix as $[\mathbf{D}_s \;\; \mathbf{D}_v]$
4. form the $\mathbf{H}$ matrix as $\begin{bmatrix} \mathbf{H}_s \\ \mathbf{H}_v \end{bmatrix}$
5. update matrix $\mathbf{H}_s$ according to (5.7a)
6. update matrix $\mathbf{D}_s$ according to (5.7b)
7. update matrix $\mathbf{H}_v$ according to (5.7c)
8. **end while**
Output: $\mathbf{D}_s$, $\mathbf{H}_s$

5.3 Experiment

5.3.1 Data

In our experiments, we employ the Noizeus speech corpus[1] [25], which is composed of 30 phonetically balanced sentences belonging to six speakers (three males and three females). The recorded speech was originally sampled at 25 kHz which were then downsampled to 8 kHz and filtered to simulate receiving frequency characteristics of telephone handsets. For our experiments, we keep the clean part of the corpus and generate a corresponding set of stimuli corrupted by additive non-stationary noise [26] (e.g., street noise) with mean SNR level $\approx$6 dB. It is worth to mention that the distributions of the 30 sentences uttered by six different speakers are as follows: the first 5 sentences (sp01, sp02, ..., sp05) belong to speaker 1, the second 5 sentences (sp06, sp07, ..., sp10) correspond to speaker 2, and so on.

[1]The Noizeus speech corpus is publicly available online at the following url: http://www.utdallas.edu/~loizou/speech/noizeus.

5.3.2 Parameters Used

In our simulation, we have used the following parameters giving the best results:

The sparsity parameter for the speech $\lambda_s = 0.15$ ($\lambda_s \in [0.05, 0.1, 0.15, 0.2]$), the sparsity parameter for the noise $\lambda_v = 0.05$ ($\lambda_v \in [0, 0.05, 0.1]$), and the sparsity parameter for the precomputed noise, $\lambda = 0.2$ ($\lambda \in [0.2, 0.3, 0.4]$). The number of precomputed noise basis functions in $\mathbf{D}_v$ is 8 and the number of speech basis functions in $\mathbf{D}_s$ is 256. Also, the number of filters for the uniform filter bank M is 64 and the length of each filter L is 256.

5.3.3 Results and Analysis

Illustrative plots are presented in Fig. 5.3 for the speech enhancement of Noizeus corpus, showing the results of the proposed method for six different speakers. It shows that the results of the proposed method are independent to the speaker.

The performance measures in terms of distortion between the noisy input signal $\mathbf{y}$ and the clean signal $\mathbf{s}$, D_{ys} $(=(\sum_i |y_i - s_i|^2)^{1/2})$ as well as the distortion between the reconstructed signal $\mathbf{y_{rec}}$ and the clean signal $\mathbf{s}$, $D_{y_{rec}s}$ $(=(\sum_i |y_{rec\,i} - s_i|^2)^{1/2})$, together with the the percentage enhancement, $E(\%)$ $(=\left(\frac{D_{ys} - D_{y_{rec}S}}{D_{ys}}\right) \times 100)$ are presented in Table 5.1. The percentage speech enhancement $E(\%)$ as achieved by different speakers is consistently higher than the results obtained by the relevant methods based on t–f masking and GSS approaches. In Fig. 5.4, the error plot for the overall results of enhancement is presented, which shows the low variations of the average results with respect to different speakers.

The results of PESQ (ITU P.862 standard) scores for different speakers are presented in Table 5.2 showing the improvement of the perceptual quality of speech signals up to 0.738.

In Fig. 5.5, the bar plot for the overall results of speech enhancement is presented, which shows the average improvements of speech with respect to different speakers and various types of nonstationary noise such as street noise, restaurant noise, and car noise. Similarly, Fig. 5.6 shows the corresponding average PESQ improvements of speech for different speakers and nonstationary noises. From the results shown in Figs. 5.5 and 5.6, it is apparent that the noise reduction and the PESQ improvement are higher for the outdoor 'street' noise than the indoor 'car' noise and 'restaurant' noise.

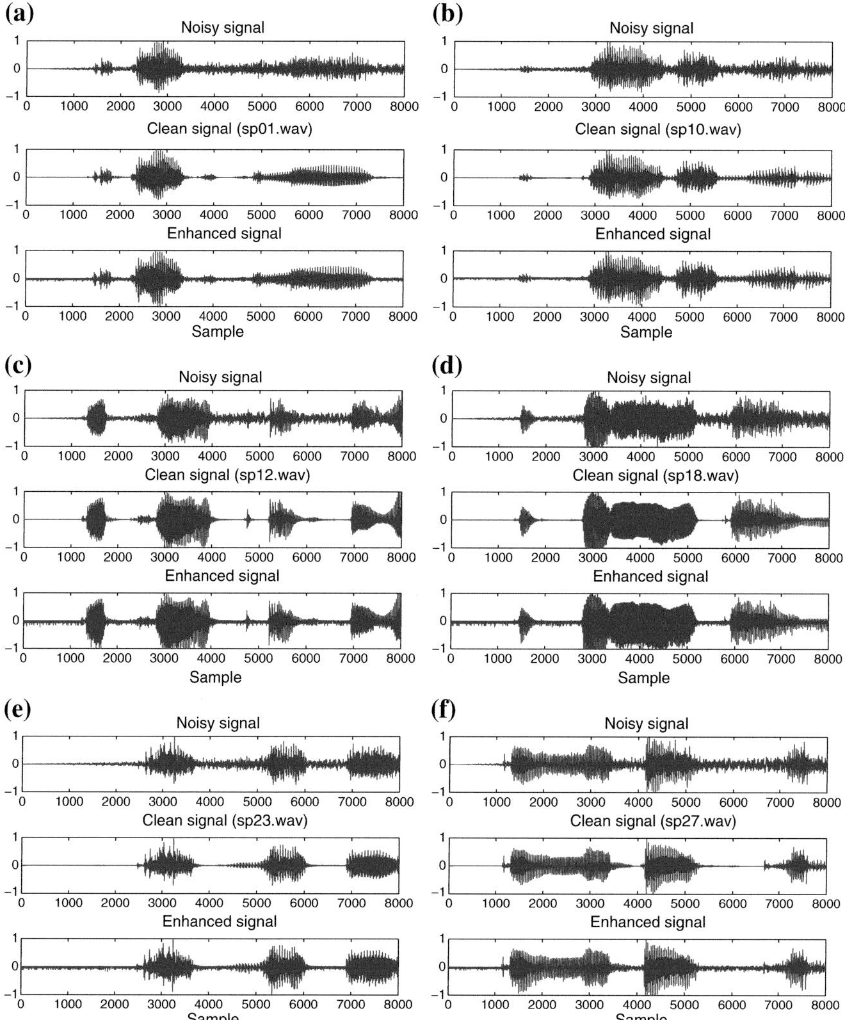

Fig. 5.3 Illustrative results of speech enhancement for the input noisy signals with nonstationary 'street noise' (SNR $\approx$ 6 dB) for six different speakers (three males and three females); **a** sp01.wav **b** sp10.wav **c** sp12.wav **d** sp18.wav **e** sp23.wav **f** sp27.wav

5.4 Discussion and Conclusion

In this study, a novel combined filter bank/MNMF scheme has been proposed for speech enhancement in real-world nonstationary noise environment. The proposed method is shown to improve the overall noise reduction by up to 30 % among different speakers and PESQ score up to 0.74 based on the experiments using recorded

Table 5.1 Performance evaluation for the Noizeus database for six different speakers (three male and three female)

File name	D_{ys}	$D_{y_{recS}}$ Proposed method	$E(\%)$ Proposed method	$D_{y_{recS}}$ t–f masking method	$E(\%)$ t–f masking method	$D_{y_{recS}}$ GSS method	$E(\%)$ GSS method
sp01	5.27	3.06	41.93	4.06	22.96	5.00	5.12
sp02	5.64	3.90	30.85	4.94	12.41	8.91	NA
sp03	4.06	3.09	23.89	3.80	6.40	4.81	NA
sp04	6.27	5.18	17.38	5.28	15.79	7.42	NA
sp05	4.54	3.29	27.53	4.15	8.59	6.05	NA
sp06	4.48	3.69	17.63	5.39	NA	4.13	7.81
sp07	4.79	3.55	25.89	4.79	0	4.75	0.84
sp08	4.06	3.50	13.79	4.20	NA	3.83	5.67
sp09	5.14	3.92	23.74	4.61	10.31	5.00	2.72
sp10	4.63	3.44	25.70	4.87	NA	4.50	2.81
sp11	3.55	2.94	17.18	4.26	NA	3.47	2.25
sp12	6.25	4.39	29.76	7.43	NA	7.68	NA
sp13	5.35	3.91	26.92	4.99	6.73	5.69	NA
sp14	5.05	2.90	42.57	3.74	25.94	3.04	39.80
sp15	7.10	5.35	24.65	5.50	22.54	9.93	NA
sp16	3.12	3.08	1.28	3.88	NA	2.89	7.37
sp17	4.62	3.75	18.83	5.63	NA	6.10	NA
sp18	7.44	5.69	23.52	7.87	NA	7.24	2.69
sp19	3.68	3.57	2.99	3.75	NA	3.41	7.34
sp20	4.86	4.24	12.76	5.68	NA	4.31	11.32
sp21	5.14	3.21	37.55	5.07	1.36	7.01	NA
sp22	6.52	4.61	29.29	6.94	NA	6.76	NA
sp23	5.18	3.65	29.54	4.46	13.90	6.21	NA
sp24	6.03	4.63	23.22	4.94	26.04	6.53	NA
sp25	4.36	3.53	19.04	4.56	NA	4.36	0
sp26	3.84	2.79	27.34	3.80	1.04	6.31	NA
sp27	6.58	4.07	35.15	6.20	5.78	6.42	2.43
sp28	3.82	2.95	22.77	4.03	NA	3.89	NA
sp29	2.78	2.46	11.51	3.41	NA	4.09	NA
sp30	4.08	3.02	25.98	4.78	NA	4.18	NA

D_{ys} The distortion between the noisy and clean signals; $D_{y_{recS}}$ The distortion between the reconstructed and clean signals; $E(\%)$ The percentage of enhancement; *NA* Not Applicable (i.e. not working)

speech data in real nonstationary noise environments. It has been also shown that the proposed method outperforms the relevant methods based on GSS and a posteriori SNR-based time–frequency (t–f) approach.

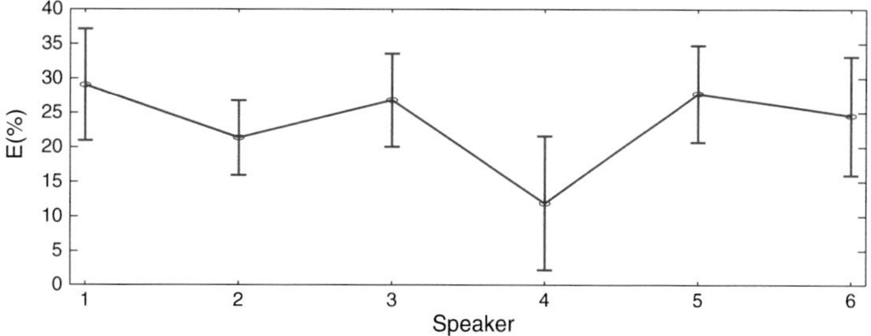

Fig. 5.4 The error plot of average enhancement (%) for the proposed method with respect to six different speakers. Bar corresponds to the standard error from the mean

Table 5.2 The improvement of PESQ (perceptual evaluation of speech quality) score

	File	Improved PESQ	Mean improved PESQ
Speaker 1	sp01	0.520	0.405
	sp02	0.363	
	sp03	0.418	
	sp04	0.245	
	sp05	0.481	
Speaker 2	sp06	0.411	0.455
	sp07	0.515	
	sp08	0.300	
	sp09	0.531	
	sp10	0.520	
Speaker 3	sp11	0.417	0.344
	sp12	0.345	
	sp13	0.363	
	sp14	0.092	
	sp15	0.504	
Speaker 4	sp16	0.580	0.465
	sp17	0.549	
	sp18	0.630	
	sp19	0.142	
	sp20	0.426	
Speaker 5	sp21	0.452	0.443
	sp22	0.613	
	sp23	0.336	
	sp24	0.548	
	sp25	0.270	
Speaker 6	sp26	0.738	0.483
	sp27	0.419	
	sp28	0.287	
	sp29	0.527	
	sp30	0.444	

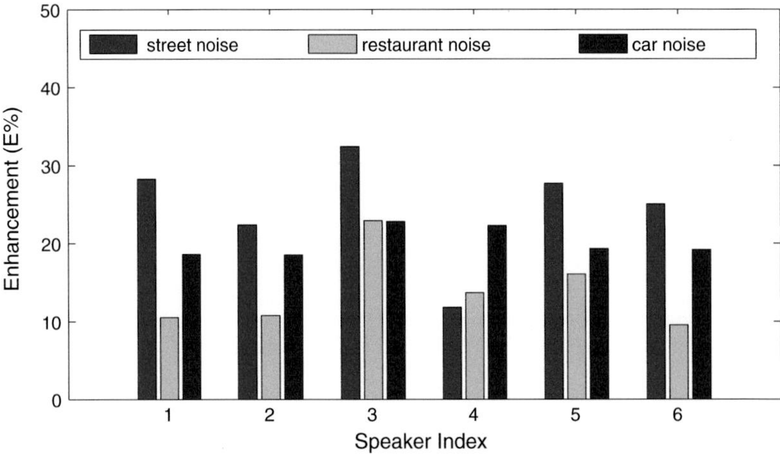

Fig. 5.5 The bar plot for average speech enhancement (E%) of the proposed method with respect to six different speakers and different types of nonstationary noise

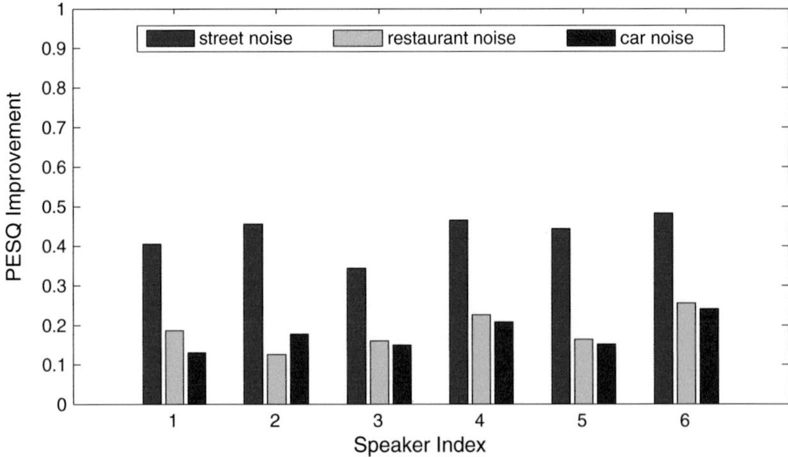

Fig. 5.6 The bar plot for average PESQ improvement of the proposed method with respect to six different speakers and different types of nonstationary noise

The choice of filter banks makes the proposed method flexible and usable for a wide variety of input signals. For an input signal other than speech signal, such as transient signal, we could consider a nonuniform filter bank having filters with unequal bandwidth. Moreover, there is no need to concern about the phase information or inverse transform with the proposed method unlike other methods based on spectrogram/short-term Fourier transform (STFT). Also, each subband signal would be simpler and sparser than the noisy input signal. That is why the proposed method is less ambiguous and less complex than the spectro-temporal based methods.

However, there remains some issues to be addressed for future work, which include (1) detail evaluation of the method in various real-world nonstationary noise environments using both objective and subjective measures, (2) performance improvement by adaptively selecting the number of basis functions for the precomputed noise dictionary learning based on length of noise-only segments, and (3) optimal parameters selection for the modified NMF scheme to enhance the performance.

References

1. Y. Hu, P.C. Loizou, Subjective comparison of speech enhancement algorithms, in *Proceedings ICASSP'06*, vol. 1 (2006), pp. 153–156
2. E. Hansler, G. Schmidt, *Topics in Acoustic Echo and Noise Control* (Springer, Berlin, 2006)
3. Y. Ephraim, D. Malah, Speech enhancement using a minimum mean-square error short-time spectral amplitude estimator, in *IEEE Transactions on Audio, Speech, and Language Processing*, vol. ASSP-32 (1984), pp. 1109–1121
4. N. Virag, Single channel speech enhancement based on masking properties of the human auditory system. IEEE Trans. Speech Audio Process. **7**(2), 497–513 (1997)
5. D.E. Tsoukalas, J.N. Mourjopoulos, G. Kokkinakis, Speech enhancement based on audible noise suppression. IEEE Trans. Speech Audio Process. **5**(6), 497–513 (1997)
6. D.D. Lee, H.S. Seung, Algorithms for non-negative matrix factorization. Adv. Neural Inf. Proc. Syst. **13**, 556–562 (2001)
7. Y.-X. Wang, Y.-J. Zhang, Nonnegative matrix factorization: a comprehensive review. IEEE Trans. Knowl. Data Eng. **25**, 6 (2013)
8. A.K. Kattepur, F. Jin, F. Sattar, Single channel source separation for convolutive mixture with application to respiratory sounds, in *IEEE-EMBS Conference on Bio-inspired Systems and Signal Processing (BIOSIGNALS)* (2010)
9. B. Ghoraani, S. Krishnan, Time-frequency matrix feature extraction and classification of environmental audio signals. IEEE Trans. Audio Speech Lang. Process. **19**(7), 2197–2209 (2011)
10. M. Kim, J. Yoo, K. Kang, S. Choi, Nonnegative matrix partial co-factorization for spectral and temporal drum source separation. IEEE J. Sel. Top. Signal Process. **5**(6), 1192–1204 (2011)
11. G. Shah, P. Koch, C.B. Papadias, On the blind recovery of cardiac and respiratory sounds. IEEE J. Biomed. Health Inform. **19**(1), 151–157 (2015)
12. H. Chung, E. Plourde, B. Champagne, Regularized NMF-based speech enhancement with spectral components modeled by Gaussian mixures, in *IEEE International Workshop on Machine Learning for Signal Processing* (2014), pp. 21–24
13. H.-T. Fan, J.-W. Hung, X. Lu, S.-S. Wang, Y. Tsao, Speech enhancement using segmental nonnegative matrix factorization, in *ICASSP'14* (2014), pp. 4483–4487
14. K. Kwon, J.W. Shin, S. Sonowal, I. Choi, N.S. Kim, Speech enhancement combining statistical models and NMF with update of speech and noise bases, in *ICASSP'14* (2014)
15. T.G. Kang, K. Kwon, J.W. Shin, N.S. Kim, NMF-based target source separation using deep neural network. IEEE Signal Process. Lett. **22**, 2 (2015)
16. Y. Hu, M. Bhatnagar, P. Loizou, A cross correlation technique for enhancing speech corrupted by colored noise, in *Proceedings of the IEEE Conference on Acoustics, Speech, Signal Processing* (2001), pp. 673–676
17. S.-J. Lee, S.-H. Kim, Noise estimation based on standard deviation and sigmoid function using a posteriori signal to noise ratio in nonstationary noisy environments. Int. J. Control Autom. Syst. **6**(6), 818–827 (2008)
18. P.P. Vaidyanathan, *Multirate Systems and Filter Banks* (Prentice-Hall, Englewood Cliffs, 1993)
19. A.N. Akansu, R.A. Haddad, *Multiresolution Signal Decomposition: Transforms, Subbands, and Wavelets* (Academic Press, Orlando, 1992)

20. M.R. Portnoff, Time-frequency representation of digital signals and systems based on short-time Fourier analysis. IEEE Trans. ASSP **28**(1), 55–69 (1980)
21. P.O. Hoyer, Non-negative sparse coding, in *IEEE Workshop Neural Networks for Signal Processing* (2002), pp. 557–565
22. J. Eggert, E. Kmrner, Sparse coding and NMF, in *IEEE Conference Neural Networks*, vol. 4 (2004), pp. 2529–2533
23. C.J. Lin, Projected gradient methods for non negative matrix factorization. Neural Comput. **19**(10), 2756–2779 (2007)
24. D. Kim, S. Sra, I.S. Dhillon, Fast newton-type methods for the least squares non negative matrix approximation problem, in *SIAM Conference on Data Mining* (2007)
25. Y. Hu, P.C. Loizou, Subjective comparison and evaluation of speech enhancement algorithms. Speech Commun. **49**(78), 588–601 (2007)
26. H. Hirsch, D. Pearce, The Aurora experimental framework for the performance evaluation of speech recognition systems under noisy conditions, in *ISCA ITRW ASR2000*, France (2000)

Chapter 6
Separation of Stellar Spectra Based on Non-negativity and Parametric Modelling of Mixing Operator

Inès Meganem, Shahram Hosseini and Yannick Deville

Abstract In this chapter, we propose blind source separation methods to extract stellar spectra from hyperspectral images. The presented work particularly concerns astrophysical data cubes from the multi-unit spectroscopic explorer (MUSE) European project. These data cubes here consist of a spectrally and spatially transformed version of the light signals emitted by dense fields of stars, due to the influence of atmosphere and of the MUSE instrument, which is modelled by the so-called point spread function (PSF). The spectrum associated with each pixel of these data cubes can thus result from contributions of different star signals. We first derive the associated mixing model, taking into account the PSF properties. We then propose two separation methods based on the non-negativity of observed data, source spectra and mixing parameters. The first method is based on non-negative matrix factorization (NMF). The second one, which is the principal contribution in this chapter, uses the proposed parametric modelling of the mixing operator and performs alternate estimation steps for the star spectra and PSF parameters. Several versions of this method are proposed, in order to eventually combine good estimation accuracy and reduced computational load. Tests are performed with simulated but realistic data. The separation method based on parametric modelling gives very satisfactory performance.

I. Meganem · S. Hosseini · Y. Deville (✉)
IRAP, CNRS, Université Paul Sabatier, 14 Avenue Edouard Belin, 31400 Toulouse, France
e-mail: Yannick.Deville@irap.omp.eu

I. Meganem
e-mail: ines.meganem@yahoo.fr

S. Hosseini
e-mail: Shahram.Hosseini@irap.omp.eu

G.R. Naik (ed.), *Non-negative Matrix Factorization Techniques*,
Signals and Communication Technology, DOI 10.1007/978-3-662-48331-2_6

6.1 Introduction

In many real applications, the observed signals provided by the sensors are mixtures of some other signals. The signals composing the mixtures, called "sources", can be useful for some scientific purpose and one then needs to extract them from the observed mixtures. Blind source separation (BSS) [1, 2] is often used to this end. It consists in estimating the source signals, with limited prior information about the mixing model and the sources, where the mixing model defines the expression of the observed signals (mixtures) with respect to the source signals.

In most existing BSS methods, the mixing model is linear, instantaneous and invariant. Four main classes of BSS methods have been developed for this model [1]. These classes depend on the assumptions which are made on the sources and/or mixing parameters. The choice among these classes thus essentially depends on the data to be processed in the considered application. The oldest class corresponds to methods based on independent component analysis (ICA) [2–6]. These methods assume the statistical independence of the sources, and can thus only be used if the sources meet this condition. The methods in the second class are based on the sparsity properties, and more specifically on the "joint" sparsity properties, of the sources. This often consists in isolating each source in a certain zone of the signals. This class is often referred to as sparse component analysis (SCA) [1, 7–11]. The third class consists of methods based on non-negative matrix factorization (NMF), which need the non-negativity of the observations, mixing coefficients and sources [12–16]. The last class corresponds to Bayesian methods [1, 17, 18], that can use different prior information about the observations, sources or mixing coefficients. They usually need to associate probability laws with the involved components.

In this chapter, we are specially interested in separating stellar spectra from hyperspectral images. Our investigation is intended for an astrophysical instrument in particular, but the developed methods can also be useful for similar data.

Hyperspectral imaging is achieved by the acquisition of images in hundreds of contiguous spectral bands. Thus, each pixel in the scene can be associated with a high-resolution spectrum corresponding to measurements obtained at a large number of wavelengths.

The work presented here is within the framework of the Multi Unit Spectroscopic Explorer (MUSE) European project. Multi Unit Spectroscopic Explorer is a new-generation integral-field spectrograph which was mounted on the Nasmyth platform of the fourth Very Large Telescope (VLT) unit in Paranal, Chile on January 19, 2014. It will provide hyperspectral astrophysical images of about 3500 bands in the visible and near-infrared domain [465 nm, 930 nm]. These images require some processing methods to permit a better exploitation of their astrophysical content by scientists.[1]

One of these needed processing methods is BSS, which particularly concerns MUSE dense stellar field images. In these images, stars are not seen as point-like

---

[1] Our work is part of the Dedicated Algorithms for HyperspectraL Imaging in Astronomy (DAHLIA) project, funded by the French ANR agency, which aims at developing such methods.

objects but spatially spread with a certain radius. This is due to the point spread function (PSF) resulting from the instrument and atmospheric effects. As a result, each pixel in the observed image can contain contributions from several stars. Each star is characterised by a spectrum. We thus aim at separating the mixtures of the star spectra present in a hyperspectral image to permit their study by astrophysicists. In this chapter, we propose new BSS methods to achieve this goal.

As will be shown in Sect. 6.2, the considered mixing model is linear but spectrally variant, which makes the use of some classical BSS methods difficult. We are also facing highly correlated sources: in dense stellar fields, stars have similar ages and thus have very similar spectra, essentially concerning absorption lines. This excludes the use of ICA. Besides, because of the PSF spatial spreading effect, the joint-sparsity-based BSS methods are not suitable, if we consider the data spatially. And spectrally, even if the continuum of the spectra is removed, the resulting signals still contain absorption lines located at the same wavelengths, which makes the joint-sparsity assumption difficult to meet. On the contrary, the non-negativity condition is verified by all the data (including sources and mixing parameters). We thus chose to use methods based on this assumption, starting with NMF. As will be detailed in Sect. 6.3, NMF, which is a *blind* method, has been largely used to analyse hyperspectral data. However, it generally does not guarantee a unique solution and is very sensitive to initialisation.

When the mixing model is partially known, *semi-blind* approaches may also be used. Then, in addition to NMF, we propose a second separation approach also exploiting non-negativity, but using a parametric model for the mixing coefficients. This method alternately performs estimation steps for the star spectra, using a least squares algorithm with non-negativity constraints, and for the mixing parameters by a projected gradient descent algorithm. This choice of method, however, does not exclude the possible use of Bayesian methods to tackle this astrophysical problem. A recent work using the Bayesian framework, also done for MUSE project, has been presented in [19].

In Sect. 6.2, we develop the considered mixing model and also show how it can be parametrised. Then, Sect. 6.3 presents our first source separation method based on NMF. In Sect. 6.4, we introduce several versions of our second separation method, which is the core of this chapter. In Sect. 6.5, we report on test results obtained with realistic simulated data.

6.2 Data Modelling

6.2.1 Assumptions About the PSF and the Data

The MUSE PSF depends on both spatial position and wavelength. It does not satisfy the usual invariance property with respect to spatial and spectral shifts. For a hypothetical observed object, which is point-like in the spectral and spatial domains and located at the spatiospectral point (z, μ), the value of its related PSF at the spatiospectral point (p, λ) can be denoted by $h^{PSF}_{z,\mu}(p, \lambda)$, where z and p are 2D spatial coordinates, μ and λ are spectral coordinates. To derive a simplified mixing model, we use the following assumptions concerning the MUSE PSF:

- $\mathcal{A}$**1**: The PSF is separable into the spatial PSF, called the field spread function (FSF), and the spectral PSF, called the line spread function (LSF):

$$h^{PSF}_{z,\mu}(p, \lambda) = h^{FSF}_{z,\mu}(p)\, h^{LSF}_{z,\mu}(\lambda).$$

- $\mathcal{A}$**2**: The LSF $h^{LSF}_{z,\mu}(\lambda)$ only spreads over a few spectral samples around μ, i.e. its values are negligible at other wavelengths λ.
- $\mathcal{A}$**3**: The FSF $h^{FSF}_{z,\mu}(p)$ slowly changes spectrally , i.e. with respect to μ.

These assumptions result from a study [20] performed within the framework of the MUSE project, and they are "realistic" (more precisely, Assumption $\mathcal{A}$**1** is a required approximation, which is widely used in the MUSE project for the sake of simplicity).

In our work, we also assume that the sky spectrum has been removed from the observed data by the preprocessing stages available in the MUSE project. Moreover, as realistic prior information, the star locations are considered to be known, thanks to Hubble Space Telescope[2] images.

6.2.2 Mixing Model

We now show how we derive the mixing model, using the previous assumptions. Stars are supposed to be spatially point-like objects in the sky before the PSF effect. The spectrum of a star i located at the 2D spatial position z_i is denoted by $e_i(\lambda)$. Let $e_i(\lambda)\delta(p - z_i)$ be the contribution of this star at the 2D spatial position p (where δ is the Dirac impulse). After taking into account the PSF effect, the contribution of this star in the observed data can be expressed as

$$\begin{aligned} y_{z_i}(p, \lambda) &= \iint e_i(\mu)\delta(z - z_i) h^{PSF}_{z,\mu}(p, \lambda) dz d\mu \\ &= \int_{\mathbb{R}} e_i(\mu) h^{PSF}_{z_i,\mu}(p, \lambda) d\mu. \end{aligned} \tag{6.1}$$

Assuming that the PSF is separable into the FSF and the LSF (Assumption $\mathcal{A}$**1**), we have

$$y_{z_i}(p, \lambda) = \int_{\mathbb{R}} e_i(\mu) h^{FSF}_{z_i,\mu}(p) h^{LSF}_{z_i,\mu}(\lambda) d\mu. \tag{6.2}$$

[2]Hubble is a spatial telescope in orbit around the Earth since 1990. It is coupled with many spectrometers. This permits it to cover a spectral domain spreading from the infra-red to near ultra-violet wavelengths.

Thanks to Assumption $\mathcal{A}\mathbf{2}$, the LSF values $h^{LSF}_{z_i,\mu}(\lambda)$ for the selected wavelength λ can be considered to be equal to zero outside a small interval of values of μ with width $2K$ and centred on λ. This yields

$$y_{z_i}(p,\lambda) = \int\limits_{\lambda-K}^{\lambda+K} e_i(\mu) h^{FSF}_{z_i,\mu}(p) h^{LSF}_{z_i,\mu}(\lambda) d\mu. \tag{6.3}$$

Now using Assumption $\mathcal{A}\mathbf{3}$, yields $h^{FSF}_{z_i,\mu}(p) \simeq h^{FSF}_{z_i,\lambda}(p)$ for $\mu \in [\lambda - K, \lambda + K]$. We thus obtain

$$y_{z_i}(p,\lambda) = h^{FSF}_{z_i,\lambda}(p) \int\limits_{\lambda-K}^{\lambda+K} e_i(\mu) h^{LSF}_{z_i,\mu}(\lambda) d\mu. \tag{6.4}$$

When a pixel contains contributions from S stars, its overall value at each wavelength reads[3]

$$\begin{aligned} y(p,\lambda) = \sum_{i=1}^{S} y_{z_i}(p,\lambda) &= \sum_{i=1}^{S} h^{FSF}_{z_i,\lambda}(p) \int\limits_{\mathbb{R}} e_i(\mu) h^{LSF}_{z_i,\mu}(\lambda) d\mu \\ &= \sum_{i=1}^{S} m_i(p,\lambda) x_i(\lambda) \end{aligned} \tag{6.5}$$

where

$$x_i(\lambda) = \int\limits_{\mathbb{R}} e_i(\mu) h^{LSF}_{z_i,\mu}(\lambda) d\mu \tag{6.6}$$

is the ith star spectrum spectrally convolved by the LSF[4] and

$$m_i(p,\lambda) = h^{FSF}_{z_i,\lambda}(p). \tag{6.7}$$

For any wavelength λ, (6.5) yields in matrix form

$$\boldsymbol{y}(\lambda) = \boldsymbol{M}(\lambda)\boldsymbol{x}(\lambda) \tag{6.8}$$

[3] Integrating the LSF over the whole spectrum or just over the interval $[\lambda - K, \lambda + K]$ is the same because the LSF is equal to zero outside this interval.

[4] We call (6.6) a "convolution" but note that it is a misnomer because, due to the spectral variability of the PSF, this integral does not correspond to the usual definition of a convolution.

where

- $\boldsymbol{y}(\lambda) = [y(p_1, \lambda), \ldots, y(p_N, \lambda)]^T$ is the vectorised observed image at wavelength λ, while N is the number of pixels in the considered field and ^{T} stands for Transpose.
- $\boldsymbol{x}(\lambda) = [x_1(\lambda), \ldots, x_S(\lambda)]^T$.
- $\boldsymbol{m_i}(\lambda) = [m_i(p_1, \lambda), \ldots m_i(p_N, \lambda)]^T$ is the vectorised FSF for star i.
- $\boldsymbol{M}(\lambda) = [\boldsymbol{m_1}(\lambda) \ldots \boldsymbol{m_S}(\lambda)]$.

The obtained model (6.8) is thus linear and "instantaneous" with respect to the spectral variable λ. However, it is spectrally variant, because the mixing matrix $\boldsymbol{M}(\lambda)$ depends on λ.

In reality, the observed data are noisy. The noise is spatially and spectrally non-stationary and can be approximated by an additive Gaussian, centred, spatially and spectrally independent process. The entire model, omitting λ in the following notations for the sake of simplicity, can thus be written for a given wavelength as

$$\boldsymbol{y} = \boldsymbol{M}\boldsymbol{x} + \boldsymbol{b} \tag{6.9}$$

with $\boldsymbol{b} \sim \mathcal{N}(0, \boldsymbol{\Gamma})$ and $\boldsymbol{\Gamma} = diag(\sigma_1^2, \ldots, \sigma_N^2)$. The noise variance depends on both pixel location and wavelength.

The noise generation is due to a Poissonian phenomenon, so its variance is an increasing function of the signal intensity at each point: the noise level is high where the signal level is high. This effect has an impact on the spectra estimation accuracy, as will be shown in Sect. 6.5.

6.2.3 Model Parametrisation

The work in [20], performed within the MUSE project, shows that the FSF can be modelled for each source i by a circular Moffat function with scale and shape parameters α and β, which depend on wavelength. Any coefficient M_{ki} of matrix $\boldsymbol{M}$ can thus be expressed as

$$M_{ki} = \frac{\beta - 1}{\pi\alpha^2}\left(1 + \frac{||p_k - z_i||_F^2}{\alpha^2}\right)^{-\beta}, \quad \alpha \geq 0,\ \beta > 1 \tag{6.10}$$

where z_i is the spatial position of source i, p_k the spatial position of the kth pixel, and $||.||_F$ is the Frobenius norm.

Therefore, for one given wavelength, the matrix $\boldsymbol{M}$ depends on the (α, β) parameters and on the star locations. It thus depends on the considered wavelength, due to α and β.

6.3 First Proposed Separation Method: LSQ-NMF

As a first step, we only consider the general mixing model (6.8), i.e. we do not yet take into account the structure (6.10) of the mixing matrix (we also disregard noise). As explained in Sect. 6.1, we exploit the non-negativity of the data (sources and mixing coefficients). This may suggest to process these data with NMF methods [12], whose principle is defined below.

Principle of NMF: Given a non-negative matrix $\boldsymbol{V}$, non-negative matrix factorization consists in finding non-negative matrix factors $\boldsymbol{W}$ and $\boldsymbol{H}$ that verify $\boldsymbol{V} \approx \boldsymbol{W}\boldsymbol{H}$.

Data involved in hyperspectral imaging being inherently non-negative, NMF can be used to analyse them. In matrix notation, each column of the matrix $\boldsymbol{V}$ in the above equation, e.g. contains the vectorised observed image at a given wavelength, and each row of this matrix corresponds to the measured spectrum in a pixel of the observed image. NMF-based methods have been largely used for hyperspectral unmixing in Earth observation [21]. In this case, each row of the matrix $\boldsymbol{H}$ in the above equation represents the spectrum of a *pure material* in the observed image, and each row of the matrix $\boldsymbol{W}$ contains the abundance fractions of all the pure materials in a given pixel of the observed image. NMF-based methods have also been used in other applications like identification of space objects [22], decomposition of audio spectra [23], extraction of features from images [24].

Unconstrained NMF generally leading to poor results, additional realistic constraints are usually employed to reduce the degrees of freedom, e.g. sum-to-one [25], minimum-volume [26], sparseness [27, 28] and smoothness [22] constraints. We should also mention the existence of other similar methods based on non-negativity of data like non-negative tensor factorization (NTF) [12, 29], non-negative matrix underapproximation (NMU) [28, 30], and alternate least square (ALS) methods [12].

NMF requires the mixing model to be linear, instantaneous *and invariant*: all observed samples, i.e. all columns of $\boldsymbol{V}$, result from the same value of the mixing matrix $\boldsymbol{W}$. On the contrary, our mixing model is variant: the column vectors $\boldsymbol{y}$ defined by (6.8), which correspond to different wavelengths, result from different values of the mixing matrix $\boldsymbol{M}$. Hereafter, we therefore propose a modified use of NMF for handling this varying mixing model.

6.3.1 Application of NMF to Considered Problem

As explained above, for NMF methods to be applicable to a set of vectors $\boldsymbol{y}$ defined by (6.8), these vectors must share the same value of the mixing matrix $\boldsymbol{M}$. Thanks to assumption $\mathcal{A}3$ (see Sect. 6.2) and (6.7), this condition is (approximately) met for all vectors $\boldsymbol{y}$ corresponding to a medium-width interval $[\lambda_1, \lambda_2]$: the FSF, and therefore the mixing matrix $\boldsymbol{M}$, smoothly vary in small intervals, so $\boldsymbol{M}$ can be considered to be constant over them. We thus can write our data, for such an interval, as

$$\boldsymbol{Y} = \boldsymbol{M}\boldsymbol{X}, \quad \text{for } \lambda \in [\lambda_1, \lambda_2], \tag{6.11}$$

with

- L: number of wavelengths in the interval
- $\boldsymbol{Y} = [\boldsymbol{y}(\lambda_1)\ldots\boldsymbol{y}(\lambda_2)] \in \mathbb{R}^{N\times L}$
- $\boldsymbol{X} = [\boldsymbol{x}(\lambda_1)\ldots\boldsymbol{x}(\lambda_2)] \in \mathbb{R}^{S\times L}$

NMF can now successively be applied to our data for several such intervals, till the estimation of the entire spectra. The width of the intervals is chosen after studying the data (here $L = 100$).

For the sake of simplicity, we here choose to apply Lee and Seung's NMF multiplicative algorithm [14] with a cost function based on the Frobenius norm.

6.3.2 NMF Initialisation: LSQ with Partially Known FSF

Unconstrained NMF methods generally do not guarantee a unique solution: if a pair of non-negative matrices $(\boldsymbol{M}, \boldsymbol{X})$ satisfies (6.11), there may exists many equivalent solutions $(\boldsymbol{MU}, \boldsymbol{U}^{-1}\boldsymbol{X})$ for non-monomial[5] matrices $\boldsymbol{U}$ with $\boldsymbol{MU} \geq 0$ and $\boldsymbol{U}^{-1}\boldsymbol{X} \geq 0$ [12]. The uniqueness can only be guaranteed under very strict conditions which are not met in our application. References [31–34] address the uniqueness problem of NMF. The NMF methods are also very sensitive to initialisation: they can converge to local minima [12]. The initialisation methods discussed in [35] may lead to a faster convergence or a lower reconstruction error at convergence, but they generally cannot avoid the convergence toward a spurious solution. In the following, we propose an initialization method based on available prior information about the mixing model.

6.3.2.1 Prior Information About the FSF

To constrain our solution and increase the probability of convergence toward an accurate estimate, we choose to take advantage of prior information about the data which is available in our investigation. For every star, we assume we have an estimate of the FSF of MUSE (for example with 10 % error). This assumption is realistic since we assume knowing the star locations and there exist works concerning FSF estimation [20].

For the NMF algorithm, the initial matrix $\boldsymbol{M}^{\boldsymbol{0}}$ is set to the above approximation of the FSF matrix $\boldsymbol{M}$ (see (6.7)).

[5] A monomial (or generalised permutation) matrix is a matrix in each row and column of which there is exactly one non-zero element.

6.3.2.2 Application of LSQ

Using $\boldsymbol{M^0}$, we derive an approximation $\boldsymbol{X^0}$ of matrix $\boldsymbol{X}$, which is then used for NMF initialisation. Based on Eq. (6.8), an estimate of $\boldsymbol{x}(\lambda)$ can be computed for each wavelength using a least squares (LSQ) approach with non-negativity constraints. This consists in finding $\boldsymbol{x^0}(\lambda)$ that minimises the following criterion

$$\|\boldsymbol{y}(\lambda) - \boldsymbol{M}^0\boldsymbol{x}(\lambda)\|_F, \quad \text{with } \boldsymbol{x}(\lambda) \geq 0, \tag{6.12}$$

and may be done using the Matlab *lsqnonneg* function. The NMF algorithm is thus initialised by $\boldsymbol{M^0}$ and $\boldsymbol{X^0} = [\boldsymbol{x^0}(\lambda_1) \ldots \boldsymbol{x^0}(\lambda_2)]$.

As shown in Sect. 6.5, this method, first presented in [36],[6] yields limited experimental performance, because the NMF is not sufficiently constrained to converge toward the right solution. Other variants of NMF, using other distances or divergences and/or using other learning rules lead to nearly similar results. Moreover, the constraints like sparseness, smoothness, sum-to-one or minimum-volume, usually used to regularise NMF cannot be used in our application because our data do not satisfy them. Therefore, we hereafter present a more powerful method, based on the parametrised mixing model using the Moffat function, defined in Sect. 6.2.3, for the FSF.

6.4 Second Proposed Separation Method: LSQ-Grd

The above method aims at simultaneously estimating the spectra and the matrix $\boldsymbol{M}$. For each star, it therefore estimates its spectrum and all its FSF coefficients (one coefficient per pixel). Many coefficients must thus be estimated with few constraints (only non-negativity). This approach therefore yields limited experimental performance.

Thanks to the FSF modelling by the Moffat function, it is possible to reduce the number of unknown parameters. As stated in Sect. 6.2.1, we assume that we know the location of every star in the studied MUSE field. Therefore, in the model defined by (6.9) and (6.10), only the spectra and the parameters (α, β) remain unknown and have to be estimated for each wavelength. Note that the final goal is the spectra estimation. This approach permits to reduce the degrees of freedom in the model, because all the entries of the mixing matrix $\boldsymbol{M}$ can be computed from the two estimated parameters (α, β).

In the following, we first describe our method in the noiseless case. We then explain how it must be modified in the noisy case. A basic version of this method was first shortly presented in [37]. We here also present some modified versions of this method including a much faster version, and more extended test results in Sect. 6.5.

---

[6]The experimental results provided in [36] should be disregarded, since it eventually turned out that the experiments reported in [36] yield an issue.

6.4.1 Minimised Criterion in Noiseless Case

In the noiseless case, for each wavelength, the criterion to be minimised is defined as

$$J = \frac{1}{2} \parallel \boldsymbol{y} - \boldsymbol{M}\boldsymbol{x} \parallel_F^2 = \frac{1}{2}(\boldsymbol{y} - \boldsymbol{M}\boldsymbol{x})^T (\boldsymbol{y} - \boldsymbol{M}\boldsymbol{x}). \tag{6.13}$$

6.4.2 Proposed Algorithm

The proposed algorithm is iterative. The estimation of the star spectra and (α, β) is performed alternately using a least squares method with non-negativity constraints (for spectra estimation) and a projected gradient descent method with a fixed step (for the (α, β) parameters). The latter is chosen for the sake of simplicity. Our global algorithm, called "LSQ-Gr" for Least SQuares—Gradient, is presented in Algorithm 1.

Algorithm 1 LSQ-Grd (basic method)

- Initialise (α, β).

repeat

- Compute matrix $\boldsymbol{M}$, using the Moffat expression (6.10) with the current values of (α, β).
- Estimate $\boldsymbol{x}$, using a least squares method with non-negativity constraints, with $\boldsymbol{M}$ fixed:

$$\boldsymbol{x} \leftarrow argmin_x J, \qquad \text{with} \quad \boldsymbol{x} \geq 0.$$

repeat

- Update (α, β) using the following rules with a fixed positive step size ε, with $\boldsymbol{x}$ fixed:

$$\alpha \leftarrow \left[\alpha - \varepsilon \frac{\partial J}{\partial \alpha}\right]_{P+}$$

$$\beta \leftarrow \left[\beta - \varepsilon \frac{\partial J}{\partial \beta}\right]_{P+}$$

where $[u]_{P+}$ corresponds to the projection of u on the interval P^+.

until gradient descent convergence

until overall algorithm convergence

The gradient descent is here stated to be "projected" since, for each iteration, the values of (α, β) are projected on the interval $P^+ \subset \mathbb{R}^+$ in order to fulfil the non-negativity constraint met by α and β and to exploit available information about their possible values. In practice, if the estimated value is outside P^+, it is replaced by the value of the nearest bound of P^+.

For our studied data, α and β vary between 1 and 2.2. We thus chose an interval $P^+ = [1, 5]$, which satisfies this condition without being too constrained.

The algorithm principle is the same in a noiseless or noisy case, only the criterion expression J changes (see Sect. 6.4.4).

6.4.3 Gradient Calculation for (α, β) Estimation in Noiseless Case

J is rewritten in scalar form to compute its gradient:

$$J = \frac{1}{2}\sum_{k=1}^{N}(y_k - (\boldsymbol{Mx})_k)^2 = \frac{1}{2}\sum_{k=1}^{N}\left(y_k - \sum_{i=1}^{S} M_{ki}x_i\right)^2 \tag{6.14}$$

with $(\boldsymbol{Mx})_k$ element k of vector $\boldsymbol{Mx}$, M_{ki} element (k, i) of matrix $\boldsymbol{M}$, y_k element k of vector $\boldsymbol{y}$ and x_i element i of vector $\boldsymbol{x}$. As shown in the appendix, the derivatives of J with respect to α and β are

$$\frac{\partial J}{\partial \alpha} = -\left[\frac{\partial \boldsymbol{M}}{\partial \boldsymbol{\alpha}}\boldsymbol{x}\right]^T (\boldsymbol{y} - \boldsymbol{Mx}), \tag{6.15}$$

and

$$\frac{\partial J}{\partial \beta} = -\left[\frac{\partial \boldsymbol{M}}{\partial \boldsymbol{\beta}}\boldsymbol{x}\right]^T (\boldsymbol{y} - \boldsymbol{Mx}), \tag{6.16}$$

with

$$\frac{\partial M_{ki}}{\partial \alpha} = \frac{2(\beta - 1)}{\pi\alpha^3}\left(1 + \frac{Z}{\alpha^2}\right)^{-\beta-1}\left(\frac{\beta - 1}{\alpha^2}Z - 1\right), \tag{6.17}$$

and

$$\frac{\partial M_{ki}}{\partial \beta} = \frac{1}{\pi\alpha^2}\left(1 + \frac{Z}{\alpha^2}\right)^{-\beta} \times \left(1 - (\beta - 1)\ln\left(1 + \frac{Z}{\alpha^2}\right)\right), \tag{6.18}$$

where $Z = ||p_k - z_i||_F^2$ (see Eq. (6.10)).

6.4.4 Noisy Case

As stated in Sect. 6.2.2, MUSE data are noisy. The noise is Gaussian, and its variance depends on both wavelength and spatial location. For our method, we take into account the noise covariance matrix for each wavelength, according to the complete mixing model given in (6.9). The minimised criterion then reads

$$J = \frac{1}{2} \parallel \boldsymbol{y} - \boldsymbol{Mx} \parallel_W^2 = \frac{1}{2}(\boldsymbol{y} - \boldsymbol{Mx})^T \boldsymbol{W}(\boldsymbol{y} - \boldsymbol{Mx}) \tag{6.19}$$

with $W = (\hat{\boldsymbol{\Gamma}})^{-1}$ and $\hat{\boldsymbol{\Gamma}}$ an estimate of the diagonal noise covariance matrix $\boldsymbol{\Gamma}$ available from the MUSE project. We thus have

$$\frac{\partial J}{\partial M_{ki}} = \left[-\boldsymbol{W}(\boldsymbol{y} - \boldsymbol{M}\boldsymbol{x})\boldsymbol{x}^T \right]_{ki}. \tag{6.20}$$

We finally obtain

$$\frac{\partial J}{\partial \alpha} = -\left[\frac{\partial \boldsymbol{M}}{\partial \alpha} \boldsymbol{x} \right]^T \boldsymbol{W}(\boldsymbol{y} - \boldsymbol{M}\boldsymbol{x}) \tag{6.21}$$

$$\frac{\partial J}{\partial \beta} = -\left[\frac{\partial \boldsymbol{M}}{\partial \beta} \boldsymbol{x} \right]^T \boldsymbol{W}(\boldsymbol{y} - \boldsymbol{M}\boldsymbol{x}). \tag{6.22}$$

For noisy data, the criterion can also be written as

$$J = \frac{1}{2} \parallel \tilde{\boldsymbol{y}} - \tilde{\boldsymbol{M}}\boldsymbol{x} \parallel_F^2, \tag{6.23}$$

with $\tilde{\boldsymbol{y}} = \boldsymbol{W}^{\frac{1}{2}}\boldsymbol{y}$ and $\tilde{\boldsymbol{M}} = \boldsymbol{W}^{\frac{1}{2}}\boldsymbol{M}$.

Thus, $\tilde{\boldsymbol{y}}$ and $\tilde{\boldsymbol{M}}$ can be used as the input arguments of the Matlab *lsqnonneg* function in the noisy case, to perform the LSQ stage of the algorithm.

6.4.5 Initialisation in Noiseless and Noisy Cases

In a basic approach, the above noiseless or noisy LSQ-Grd algorithm is separately applied to each wavelength. The intervals of possible values of α and β are assumed to be known, and these parameters are randomly initialised in these intervals for each run of LSQ-Grd.

In an improved approach, these random initialisations of α and β are only used for the first considered wavelength, which is the lowest one. Then, the other wavelengths are considered in increasing order and, for each of them, the initial values of α and β provided to LSQ-Grd are set to the estimates of α and β derived by LSQ-Grd for the *previous* wavelength. This is relevant because α and β smoothly vary spectrally. This improved algorithm, which aims at accelerating the convergence of LSQ-Grd, and takes advantage of Sequential calls of LSQ-Grd, is called S-LSQ-Grd.

The explicit initialisation of the spectra $\boldsymbol{x}$ is not necessary, since it is handled by the Matlab *lsqnonneg* function used here for the LSQ stage (4) of our LSQ-Grd algorithm, which estimates $\boldsymbol{x}$.

6.4.6 Accelerated Algorithm for All Wavelengths

The above algorithm yields high computational loads for large data sets such as those faced in the MUSE project. We therefore here propose an accelerated version of that method, which further takes advantage of the fact that the (α, β) parameters smoothly vary with respect to wavelength.

The main idea is to only apply the previous LSQ-Grd algorithm every K wavelengths (with K, e.g. set to 100) in order to estimate some equally spaced values of the (α, β) parameters, and then to interpolate between them to obtain (α, β) at each wavelength. The interpolation may be done using splines or polynomial functions.

We then just need to use LSQ to estimate the values of the spectra at all wavelengths. This spectral-interpolation version of our method based on LSQ-Grd is called SI-LSQ-Grd and is shown in Algorithm 2.

Algorithm 2 SI-LSQ-Grd

- Estimate an equally spaced subset of values of (α, β) by applying The LSQ-Grd algorithm every K wavelengths.
- Interpolate (α, β) values to obtain an estimate of (α, β) values for all the wavelengths of MUSE spectrum.
- Estimate spectra for all wavelengths from the (α, β) values, by using least squares with non-negativity constraints.

This SI-LSQ-Grd algorithm permits a considerable computing time saving as compared with the approaches described in Sect. 6.4.5, since it calls the complete LSQ-Grd algorithm K times less than in the latter approaches.

6.5 Test Results

This section presents the results obtained with the LSQ-NMF, S-LSQ-Grd and SI-LSQ-Grd algorithms. All methods were tested over simulated but realistic data provided in the framework of the MUSE project (note that real data are not available yet).

6.5.1 Data Description

Figure 6.1 shows the 16×16-pixel studied field (for one spectral band). It corresponds to the grey-level square in the image centre. Each pixel represents a 0.2×0.2 arcsec^2 region.

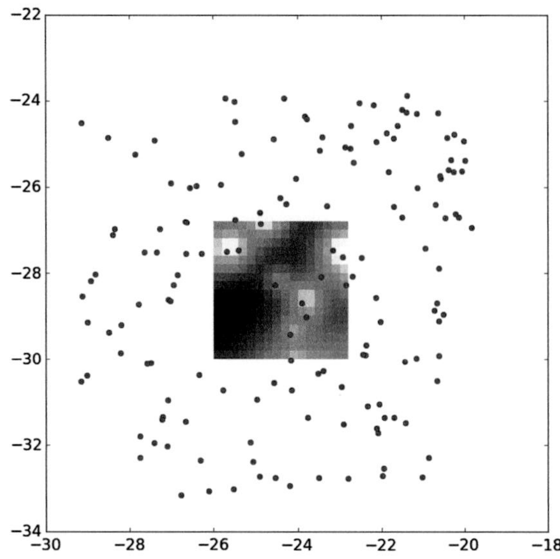

Fig. 6.1 Studied field (the *square* in the centre) and its neighbourhood

The dots show the actual star positions within the field (11 stars) and in its neighbourhood. As shown in Sect. 6.5.3, stars in the neighbourhood contribute to information inside the field. The cube corresponding to this field contains 3578 wavelengths from 465 to 930 nm, with a 0.13 nm step. For this field, the following data are available:

- A noiseless data cube and a noisy one characterised by an average signal-to-noise ratio (SNR) of 28 dB.
- An estimate of the noise variance.[7]
- Necessary data to evaluate results: for each star, we have its spectrum "convolved" by the LSF (spectrum), by the FSF (image for each wavelength) and by the total PSF (cube).

6.5.2 Performance Criteria

We here present all the performance criteria used in the following sections. Note that errors for spectra and FSF only concern the stars inside the field, even if this does not appear in notations.

[7]In real operation, the variance estimates will be provided by the data reduction software (DRS) of MUSE.

6.5.2.1 Single-Wavelength Estimation

To evaluate the performance over all wavelengths, obtained by the S-LSQ-Grd and SI-LSQ-Grd methods which estimate spectra and FSF separately at each wavelength, we use the following criteria, where "$\widehat{\ }$" corresponds to estimates:

- The relative errors computed over the entire spectra:

$$Err_{s_i} = \frac{\|\boldsymbol{s_i} - \widehat{\boldsymbol{s_i}}\|_F}{\|\boldsymbol{s_i}\|_F} \tag{6.24}$$

  where $\boldsymbol{s_i}$ is the entire spectrum of star i convolved by LSF. We also use its mean over all stars : $Err_s = mean_i(Err_{s_i})$.
- The relative error for the FSF:

$$Err_M = mean_\lambda\left(\frac{\|\boldsymbol{M} - \widehat{\boldsymbol{M}}\|_F}{\|\boldsymbol{M}\|_F}\right). \tag{6.25}$$

- The total (i.e. considering both spectra and FSF) relative reconstruction error:

$$Err_{tot} = mean_\lambda\left(\frac{\|\boldsymbol{y} - \widehat{\boldsymbol{M}}\widehat{\boldsymbol{x}}\|_F}{\|\boldsymbol{y}\|_F}\right). \tag{6.26}$$

6.5.2.2 Estimation over Spectral Intervals

For the LSQ-NMF method, which estimates spectra and FSF simultaneously over all wavelengths situated in a spectral interval, we use the following expression for the total relative reconstruction error, to essentially check the NMF convergence

$$Err_{tot} = mean_{intervals}\left(\frac{\|\boldsymbol{Y} - \widehat{\boldsymbol{M}}\widehat{\boldsymbol{X}}\|_F}{\|\boldsymbol{Y}\|_F}\right) \tag{6.27}$$

We also need the relative estimation error for spectra, which is computed as follows:

$$Err_X = mean_{intervals}\left(\frac{\|\boldsymbol{X} - \widehat{\boldsymbol{X}}\|_F}{\|\boldsymbol{X}\|_F}\right) \tag{6.28}$$

The notation "$\widehat{\ }$" here corresponds to the estimates provided by LSQ-NMF.

6.5.3 Preliminary Data Analysis

The aim of this section is to emphasise an important characteristic of the data. We here assume we exactly know the FSF and only the spectra are estimated using least squares with non-negativity constraints, wavelength by wavelength. Only the noiseless case is treated here.

Because of the spatial spread of the FSF, stars in the neighbourhood of the studied field can contribute to the information inside the field. Figure 6.2 shows the reconstruction error (Err_{tot} of Eq. (6.26)) and the estimation error for spectra (Err_S) inside the field, as functions of the "radius of neighbourhood" taken into account for the estimation. What is called "radius of neighbourhood" corresponds to a distance added to each horizontal or vertical side of our studied field, which yields an extended square field. We thus consider that all stars included in the extended field contribute to the information inside the initial field. In practice, this means that, in Eq. (6.8), we include stars situated in the fixed neighbourhood in the set of S stars involved in matrix $\boldsymbol{M}$ and vector $\boldsymbol{x}$, without changing $\boldsymbol{y}$ which only contains pixels of the initial field. However, at the end, we are only interested in the estimation of the spectra of the stars inside the studied field. Figure 6.2 thus shows that taking into account the stars in the neighbourhood of the studied field permits a much more accurate estimation of the spectra of the stars inside the field.

Table 6.1 shows, as an indication, the number of stars in the extended field, depending on the considered radius of neighbourhood.

Simulations reported in all the following sections are performed using a fixed radius of neighbourhood of 1.4 arcsec (equivalent to 7 pixels), which seems, according to Fig. 6.2, to be sufficient.

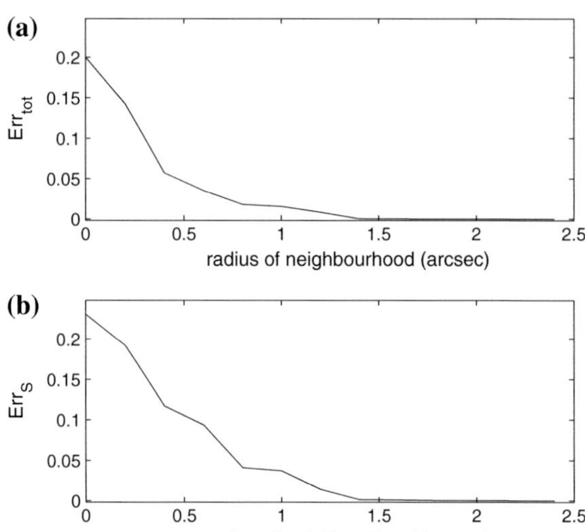

Fig. 6.2 Err_{tot} and Err_S as functions of the radius of neighbourhood taken into account for estimation—spectra are estimated with a completely known FSF—noiseless data. **a** Total error. **b** Spectra estimation error

Table 6.1 Number of stars in extended field, versus considered radius of neighbourhood

Radius (arcsec)	0	0.2	0.6	1	**1.4**	1.8	2.2
Number of stars	11	14	27	42	**59**	76	95

6.5.4 Performance of the LSQ-NMF Method

As stated in Sect. 6.3.2, for the LSQ-NMF method, we assume knowing a rough estimate of the FSF (e.g. with 10 % error). To test the method, the FSF matrix used to initialise the NMF is here generated as follows. We use the FSF values we have with the data,[8] after adding a relative noise to simulate the 10 % error. The error is then modelled by a Gaussian zero-mean additive noise with a standard deviation equal to 10 % of the FSF amplitude in the considered pixel.

Since the error over the FSF is modelled by a random noise, the LSQ-NMF method is tested for 30 different initialisations. Tests are here made over the noiseless data cube, to evaluate the method performance regardless of the noise.

Figure 6.3 shows, for one spectral interval (from the 601st to 700th wavelengths), the true spectra (thin lines) and, for one initialisation case (among the 30 cases), the spectra estimated by LSQ-NMF (thick lines). Some estimates are shifted compared to the real spectra, and this shift can vary from a spectral interval to another, which may give deformed spectra.

In these tests, the LSQ-NMF algorithm converges to solutions which yield low total errors, that is about 0.5 % in all spectral intervals. However, LSQ-NMF does not improve the error for spectra, as compared to the spectra directly estimated using LSQ from the initial estimate of FSF. The mean estimation error for the spectra is 15 %, which is unacceptably high. This limitation may be explained by the well-known non-uniqueness of the NMF solution without constraint as explained at the end of Sect. 6.3.

6.5.5 Performance of Methods Based on LSQ-Grd

6.5.5.1 S-LSQ-Grd Method

Noiseless Data

We first present results obtained with the noiseless data, to then better evaluate the noise impact.

Spectra are here estimated with a mean error Err_S of 0.8 %. The spectra estimation is thus almost perfect. This can be verified with Fig. 6.4 where it can be seen that the

[8] The FSF matrix to which we add the noise, and which is used for initialisation, is in fact an average matrix of the L true matrices of FSF corresponding to the considered spectral interval, since this FSF matrix is assumed to be constant in the interval by our LSQ-NMF method, but in reality it depends on the wavelength.

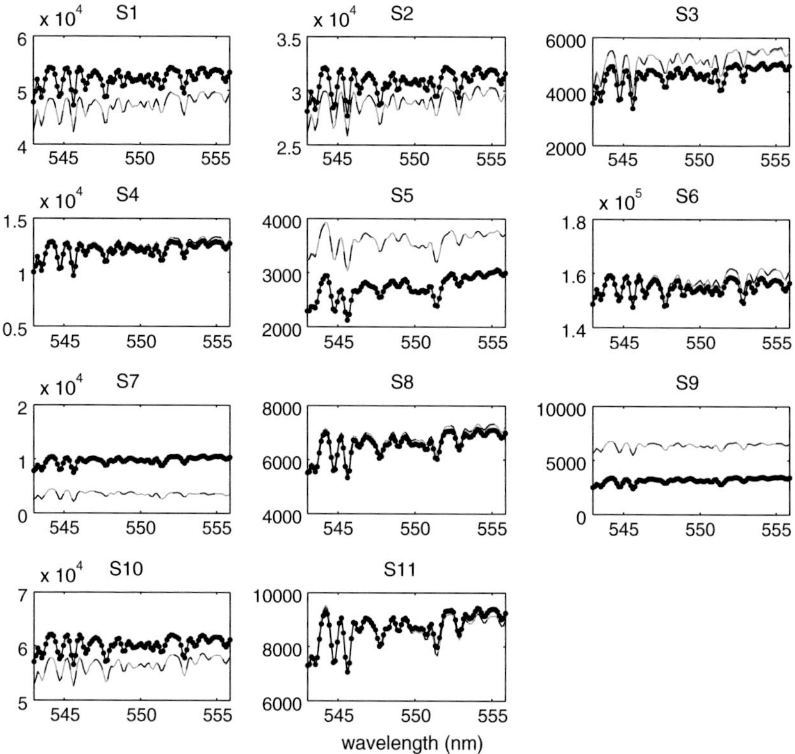

Fig. 6.3 Star spectra estimated by LSQ-NMF (*thick lines*) and true spectra (*thin lines*), for spectral interval between the 601st and 700th wavelengths

real spectra (black) and their estimates (grey) are almost perfectly superimposed. The error over the FSF is $Err_M = 1.4\,\%$, so the (α, β) estimation is very satisfactory. The reference values of α and β (solid lines) and their estimated values (dash-dotted lines) can be seen in Fig. 6.5. The total error is also small ($Err_{tot} < 1\,\%$), which shows that the algorithm converges. The noiseless case is not realistic for our application, although it shows that our algorithm gives very good results and converges to the right solution in ideal conditions.

Noisy Data

We now process the more realistic data, that is the noisy ones. Figure 6.6 shows the estimated values of (α, β) (in grey) and their true values (black).[9] Although the variations of the estimates of (α, β) versus wavelength are less smooth than in the noiseless case, the resulting overall error for the FSF remains close to the value obtained in that case, with $Err_M = 1.8\,\%$. This permits us to conclude that (α, β)

[9]The removed spectral bands correspond to a very noisy zone, since it has been used for the calibration with a laser star.

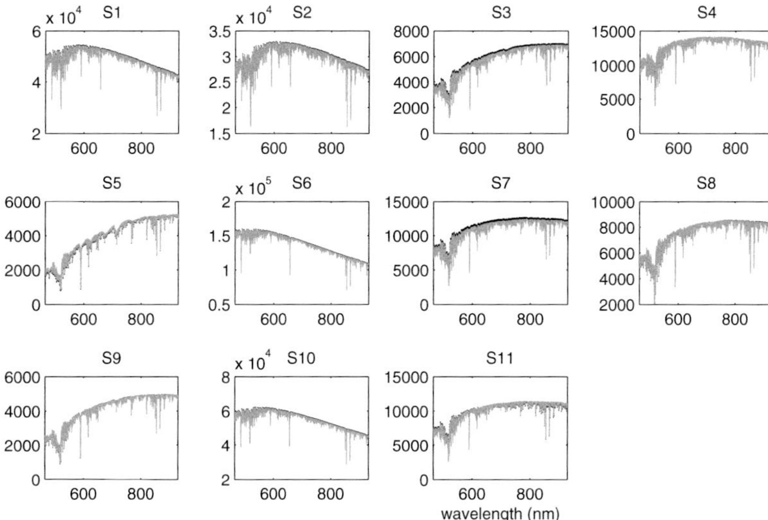

Fig. 6.4 Spectra of the 11 stars in the field: true spectra (*black*) and their estimates (*grey*)—noiseless data

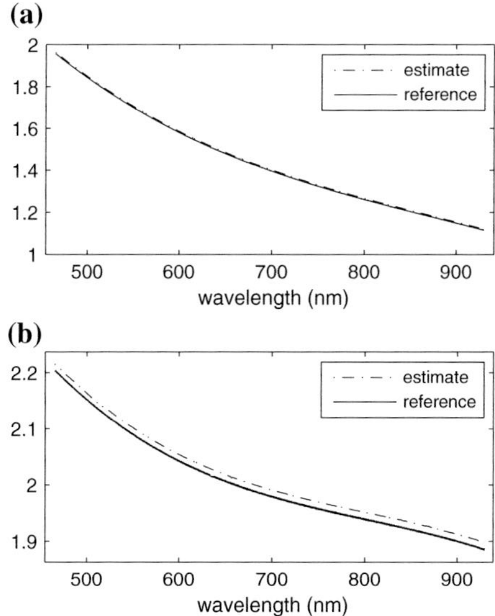

Fig. 6.5 Values of (α, β) versus wavelength: true values (*solid lines*) and their estimates (*dash-dotted lines*)—noiseless data. **a** Values of α. **b** Values of β

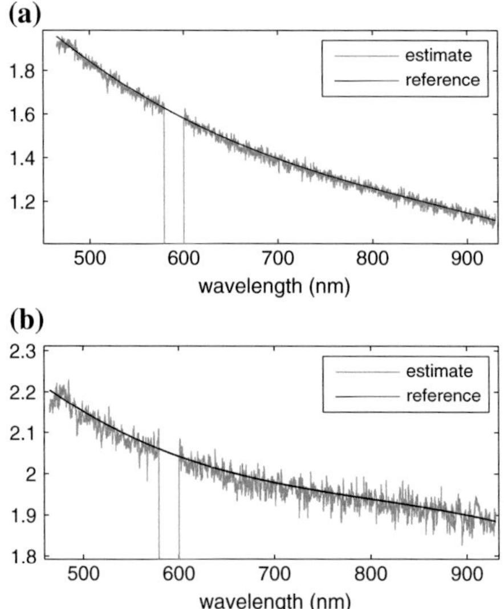

Fig. 6.6 Values of (α, β) versus wavelength: true values (*black*) and their estimates (*grey*)—noisy data. **a** Values of α. **b** Values of β

estimation is satisfactory. Concerning the spectra estimation, the mean estimation error is 7.6 %. Figure 6.7 shows the obtained errors Err_{s_i} for the 11 stars inside the field. The spectra estimation is thus less precise here and there is a disparity between the estimation qualities of the different spectra. Figure 6.8 shows the estimated spectra (grey) and the true ones (black). It can be seen that the spectra corresponding to a high estimation error in Fig. 6.7 are noisier than others.

We now analyse the disparity in spectra estimation accuracy, to understand these results. Figure 6.9 shows the star locations in the studied field. The star indices (from 1 to 11) here correspond to those in the previous figures. Here is our interpretation of results for spectra estimation in the noisy case:

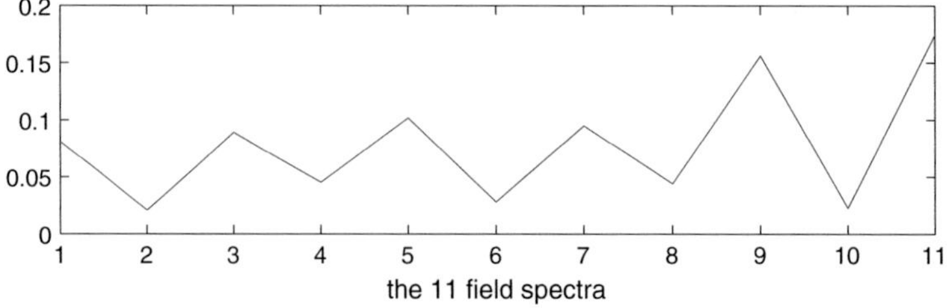

Fig. 6.7 Estimation errors for the spectra of 11 stars in the field—noisy data

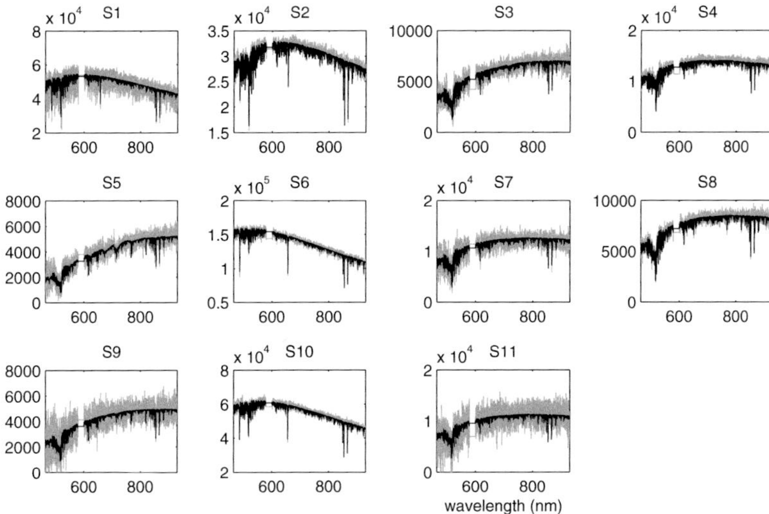

Fig. 6.8 Spectra of the 11 stars in the field, the true ones (*black*) and their estimates (*grey*)—noisy data

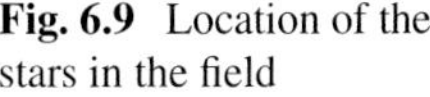

Fig. 6.9 Location of the stars in the field

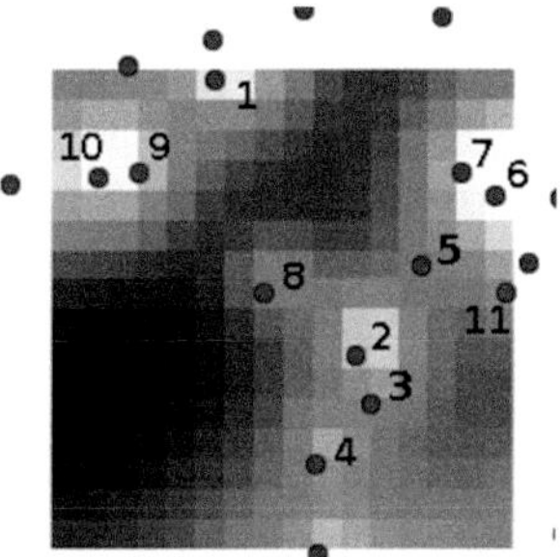

- Stars 3, 5, 7 and 9 are situated near a much brighter star (stars 2, 6, 6 and 10, respectively). The star magnitudes can be read on the axes in Fig. 6.8. As stated in Sect. 6.2.2, MUSE noise level increases with the signal level in each pixel. Around a very bright star, the noise is thus high and neighbouring stars are "drowned" in noise corresponding to a high level compared with their magnitude. As a result, the estimated spectra are noisier than others.
- Stars 1 and 11 are located in the edge of the field. The information in the field for them is thus incomplete, which can explain the high obtained errors. These stars are in fact less well estimated than others even with noiseless data.

The well-estimated spectra correspond to stars that have a high magnitude or isolated stars.

To show the performance of our method, we eventually compare the observed (unprocessed) data and our estimates. Figure 6.10 shows, for each star:

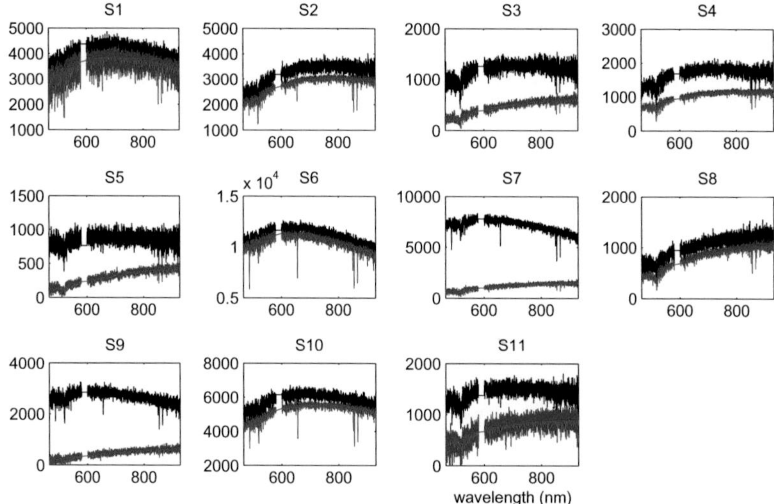

Fig. 6.10 Comparison between estimated spectra (*blue*), true spectra (*red*) and pixel spectra directly extracted from the observed cube at pixels corresponding to star location (*black*)—noisy data

- in black: the unprocessed spectrum extracted from the observed data cube at the pixel corresponding to the star location
- in red: the true star spectrum, multiplied by the true FSF coefficient corresponding to the star location. This multiplication aims at calibrating the magnitude of the true spectrum with that of the observed data (i.e. the black spectrum)
- in blue: the star estimated spectrum, multiplied by the corresponding estimated FSF coefficient.

It can be seen that even for the noisiest estimated spectra, the estimation results are much closer to the true spectra than the unprocessed data. The method permits one to obtain estimated spectra that have the right magnitude and a good continuum shape. Besides, for some of the spectra, the absorption lines are well estimated. These results obtained with noisy realistic data are thus very satisfactory.

6.5.5.2 SI-LSQ-Grd Accelerated Method

This section first aims at showing that this new accelerated version of our method yields almost the same estimation accuracy as the above S-LSQ-Grd algorithm.

With the noiseless data, the result is the same as with S-LSQ-Grd: the mean estimation error for the spectra is 0.8 %. The second step of the SI-LSQ-Grd method, i.e. the interpolation of the (α, β) values, was here performed using splines.

With the noisy data, we showed above that the variations of the estimates of (α, β) versus wavelength are not as smooth as in the noiseless case (see Fig. 6.6). The spectral interpolation of (α, β) may therefore be difficult, which may yield lower performanc for the SI-LSQ-Grd algorithm than for S-LSQ-Grd. We thus investigated

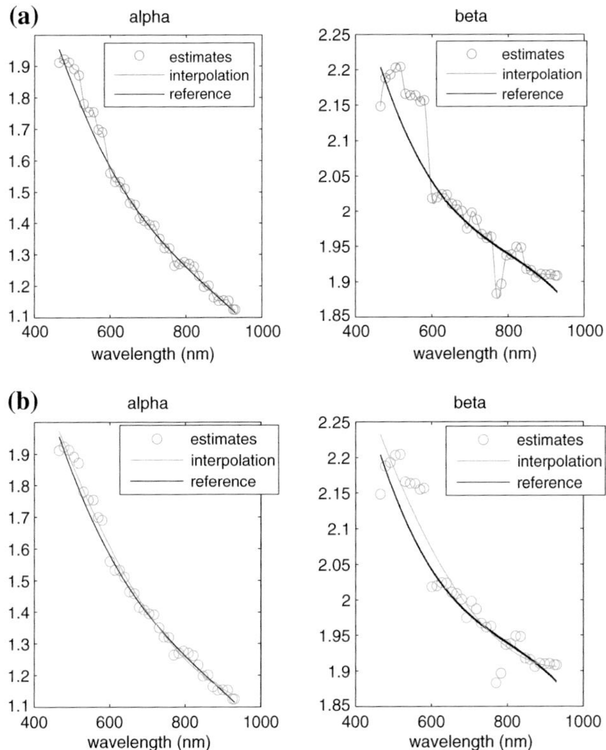

Fig. 6.11 Values of (α, β): reference values (*black*) and their estimates (*grey*)—SI-LSQ-Grd accelerated method—noisy data. **a** Interpolation with splines. **b** Interpolation with a second-order polynomial

two interpolation methods. Figure 6.11. a shows the (α, β) values estimated at the first step of the algorithm (circles) and the values obtained after interpolation with splines (grey). Figure 6.11. b shows interpolation values obtained using a second-order polynomial. In both figures, the reference values are in black. The second interpolation method gives a more satisfactory result for (α, β) values.

Concerning the spectra estimation, the first interpolation gives a mean error of 7.6 % and the second one 7.4 %. Both values are close to the value obtained with the S-LSQ-Grd method, which was 7.6 %. However, if we look at the spectra shape (not presented here), it appears that the first interpolation method (splines) gives deformed spectra, especially around the wavelengths where (α, β) are very badly estimated.

On the contrary, the spectra obtained when using the interpolation with a second-order polynomial are very similar to those obtained with the S-LSQ-Grd algorithm. The result with this method is thus satisfactory.

This new SI-LSQ-Grd accelerated version of our S-LSQ-Grd method permits to obtain the same results, when choosing the appropriate interpolation method for (α, β), in its second step. And most importantly, it is almost 100 times faster than the S-LSQ-Grd version. The computing time is of 8 hours with SI-LSQ-Grd (and unacceptably high with S-LSQ-Grd) for the studied field, the computation being done with Matlab on a double core processor, with a 2.8 GHz frequency and a 4 GB RAM.

6.6 Conclusion and Future Work

In this chapter, we presented source separation methods to extract stellar spectra from MUSE dense field hyperspectral images. We first presented the mixing model that we derived using information about the MUSE PSF. Taking into account the properties of the source and observed data and of the mixing model, we then proposed two non-negativity-based methods.

The first one uses NMF combined with an initialisation with LSQ and is successively applied to small spectral intervals. It did not give satisfactory results, which can be due to the non-uniqueness of the NMF solution without enough constraints. We thus proposed a second more efficient method, which uses the parametrisation of the mixing model. This parametrisation uses the FSF model and permits to constrain the solution and reduce the number of parameters to be estimated. This new method succeeds in alternately estimating the FSF parameters and the star spectra. Tests were made on realistic simulated MUSE data, showing very satisfactory results.

As future work, we first intend to add some improvements to the method, to get rid of the fixed adaptation step in the iterative algorithms, which could accelerate the method and enhance the convergence quality. This can be achieved using for example a Newton version of the gradient descent algorithm or by using an adaptation step which varies according to some update rules from the literature.

It would also be interesting to study the sensitivity of the method to errors on the star positions. We can also assume knowing these positions with a certain precision (e.g. 10 or 20 % of pixel size), and extend the method to estimate them and increase their precision. It can then be considered that the stars outside the studied field do not necessitate the same precision, which would reduce the number of positions to be estimated.

Finally, we should emphasise that although our investigation in this chapter was intended for an astrophysical instrument, the idea used in the developed methods may also be useful in other applications where the mixing matrix can be parametrised.

Appendix

We can write

$$\frac{\partial J}{\partial \alpha} = \sum_k \sum_i \frac{\partial J}{\partial M_{ki}} \frac{\partial M_{ki}}{\partial \alpha} \tag{6.29}$$

$$\frac{\partial J}{\partial \beta} = \sum_k \sum_i \frac{\partial J}{\partial M_{ki}} \frac{\partial M_{ki}}{\partial \beta}. \tag{6.30}$$

Derivation of (6.14) with respect to matrix $\boldsymbol{M}$ gives

$$\frac{\partial J}{\partial M_{ki}} = (y_k - (\boldsymbol{M}\boldsymbol{x})_k)(-x_i) = -\left[(\boldsymbol{y} - \boldsymbol{M}\boldsymbol{x})\boldsymbol{x}^T\right]_{ki}. \tag{6.31}$$

Calculation with respect to α then reads

$$\begin{aligned}
\frac{\partial J}{\partial \alpha} &= \sum_k \sum_i (y_k - (\boldsymbol{M}\boldsymbol{x})_k)(-x_i) \times \frac{\partial M_{ki}}{\partial \alpha} \\
&= -\sum_k (y_k - (\boldsymbol{M}\boldsymbol{x})_k) \sum_i \frac{\partial M_{ki}}{\partial \alpha} x_i \\
&= -\sum_k \left[\boldsymbol{y} - \boldsymbol{M}\boldsymbol{x}\right]_k \left[\frac{\partial \boldsymbol{M}}{\partial \boldsymbol{\alpha}} \boldsymbol{x}\right]_k \\
&= -\left[\frac{\partial \boldsymbol{M}}{\partial \boldsymbol{\alpha}} \boldsymbol{x}\right]^T (\boldsymbol{y} - \boldsymbol{M}\boldsymbol{x}).
\end{aligned} \tag{6.32}$$

The same type of expression is obtained for β:

$$\frac{\partial J}{\partial \beta} = -\left[\frac{\partial \boldsymbol{M}}{\partial \boldsymbol{\beta}} \boldsymbol{x}\right]^T (\boldsymbol{y} - \boldsymbol{M}\boldsymbol{x}). \tag{6.33}$$

To compute $\frac{\partial \boldsymbol{M}}{\partial \boldsymbol{\alpha}}$ and $\frac{\partial \boldsymbol{M}}{\partial \boldsymbol{\beta}}$, we use the expression of M_{ki} as a function of (α, β), given by Eq. (6.10). To simplify notations, as the star positions do not depend on (α, β), we introduce $Z = ||p_k - z_i||_F^2$ in expression (6.10).

This gives, for α:

$$\begin{aligned}
\frac{\partial M_{ki}}{\partial \alpha} &= \frac{\beta - 1}{\pi} \times \frac{-2}{\alpha^3} \left(1 + \frac{Z}{\alpha^2}\right)^{-\beta} \\
&\quad + \frac{\beta - 1}{\pi \alpha^2} \times (-\beta) \left(1 + \frac{Z}{\alpha^2}\right)^{-\beta-1} \times \frac{-2}{\alpha^3} \times Z \\
&= \frac{2(\beta - 1)}{\pi \alpha^3} \left(1 + \frac{Z}{\alpha^2}\right)^{-\beta-1} \left(\frac{\beta}{\alpha^2} Z - \left(1 + \frac{Z}{\alpha^2}\right)\right) \\
&= \frac{2(\beta - 1)}{\pi \alpha^3} \left(1 + \frac{Z}{\alpha^2}\right)^{-\beta-1} \left(\frac{\beta - 1}{\alpha^2} Z - 1\right)
\end{aligned} \tag{6.34}$$

And calculation for β reads[10]

$$\frac{\partial M_{ki}}{\partial \beta} = \frac{1}{\pi\alpha^2}\left(1+\frac{Z}{\alpha^2}\right)^{-\beta} - \frac{\beta-1}{\pi\alpha^2}\left(1+\frac{Z}{\alpha^2}\right)^{-\beta} \times \ln\left(1+\frac{Z}{\alpha^2}\right)$$
$$= \frac{1}{\pi\alpha^2}\left(1+\frac{Z}{\alpha^2}\right)^{-\beta} \times \left(1-(\beta-1)\ln\left(1+\frac{Z}{\alpha^2}\right)\right) \quad (6.35)$$

References

1. P. Comon, C. Jutten, *Handbook of Blind Source Separation, Independent Component Analysis and Applications* (Academic Press, Oxford, 2010)
2. A. Hyvärinen, J. Karhunen, E. Oja, *Independent Component Analysis*, Adaptive and Learning Systems for Signal Processing, Communications and Control (Wiley, New York, 2001)
3. C. Jutten, J. Herault, Blind separation of sources, part I: an adaptive algorithm based on neuromimetic architecture. Signal Process. **24**(1), 1–10 (1991)
4. P. Comon, Independent component analysis, a new concept? Signal Process. **36**(3), 287–314 (1994)
5. A. Hyvärinen, Fast and robust fixed-point algorithms for independent component analysis. IEEE Trans. Neural Netw. **10**(3), 626–634 (1999)
6. M. Babaie-Zadeh, C. Jutten, A general approach for mutual information minimization and its application to blind source separation. Signal Process. **85**(5), 975–995 (2005)
7. R. Gribonval, S. Lesage, A survey of sparse component analysis for blind source separation: principles, perspectives, and new challenges, in *Proceedings of the ESANN* (2006), pp. 323–330
8. F. Abrard, Y. Deville, A time-frequency blind signal separation method applicable to underdetermined mixtures of dependent sources. Signal Process. **85**(7), 1389–1403 (2005)
9. I. Meganem, Y. Deville, M. Puigt, Blind separation methods based on correlation for sparse possibly-correlated images, in *Proceedings of IEEE International Conference ICASSP, Dallas, USA* (2010)
10. O. Yilmaz, S. Rickard, Blind separation of speech mixtures via time-frequency masking. IEEE Trans. Signal Process. **52**, 1830–1847 (2004)
11. Y. Deville, M. Puigt, Temporal and time-frequency correlation-based blind source separation methods. part I: determined and underdetermined linear instantaneous mixtures. Signal Process. **87**(3), 374–407 (2007)
12. A. Cichocki, R. Zdunek, A.H. Phan, S.-I. Amari, *Nonnegative Matrix and Tensor Factorizations: Applications to Exploratory Multi-Way Data Analysis and Blind Source Separation* (Wiley, Chichester, 2009)
13. P. Paatero, U. Tapper, Positive matrix factorization: a non-negative factor model with optimal utilization of error estimates of data values. Environmetrics **5**(2), 111–126 (1994)
14. D. D. Lee, H. S. Seung, Algorithms for non-negative matrix factorization, in *Proceedings of the NIPS*, vol. 13 (MIT Press, 2001)
15. C.-J. Lin, Projected gradient methods for nonnegative matrix factorization. Neural Comput. **19**, 2756–2779 (2007)
16. R. Zdunek, A. Cichocki, Fast nonnegative matrix factorization algorithms using projected gradient approaches for large-scale problems. Comput. Intell. Neurosci. **2008**, 3:1–3:13 (2008)
17. K.H. Knuth, A Bayesian approach to source separation, in *Proceedings of the First International Workshop on Independent Component Analysis and Signal Separation: ICA* (1999), pp. 283–288

[10]We use: $\dfrac{\partial(a^{-x})}{\partial x} = -a^{-x}\ln(a)$.

18. S. Moussaoui, D. Brie, A. Mohammad-Djafari, C. Carteret, Separation of non-negative mixture of non-negative sources using a Bayesian approach and MCMC sampling. IEEE Trans. Signal Process. **54**(11), 4133–4145 (2006)
19. A. Selloum, Y. Deville, H. Carfantan, Separation of stellar spectra from hyperspectral images using particle filtering constrained by a parametric spatial mixing model, in *Proceedings of IEEE ECMSM* (2013)
20. D. Serre, E. Villeneuve, H. Carfantan, L. Jolissaint, V. Mazet, S. Bourguignon, A. Jarno, Modeling the spatial PSF at the VLT focal plane for MUSE WFM data analysis purpose, in *SPIE Proceedings of Astronomical Telescopes and Instrumentation*
21. J.M. Bioucas-Dias, A. Plaza, N. Dobigeon, M. Parente, Q. Du, P. Gader, J. Chanussot, Hyperspectral unmixing overview: geometrical, statistical, and sparse regression-based approaches. IEEE J. Sel. Top. Appl. Earth Obs. Remote Sens. **5**(2), 354–379 (2012)
22. V.P. Pauca, J. Piper, R.J. Plemmons, Nonnegative matrix factorization for spectral data analysis. Linear Algebra Appl. **416**, 29–47 (2006)
23. C. Fevotte, N. Bertin, J.-L. Durrieu, Nonnegative matrix factorization with the itakura-saito divergence: with application to music analysis. Neural Comput. **21**, 793–830 (2009)
24. B. Gao, W.L. Woo, S.S. Dlay, Variational regularized 2-d nonnegative matrix factorization. IEEE Trans. Neural Netw. Learn. Syst. **23**(5), 703–716 (2012)
25. D.C. Heinz, C.-I. Chang, Fully constrained least squares linear spectral mixture analysis method for material quantification in hyperspectral imagery. IEEE Trans. Geosci. Remote Sens. **39**(3), 529–545 (2001)
26. L. Miao, H. Qi, Endmember extraction from highly mixed data using minimum volume constrained nonnegative matrix factorization. IEEE Trans. Geosci. Remote Sens. **45**(3), 765–777 (2007)
27. J. Rapin, J. Bobin, J.-L. Starck, Sparse and non-negative bss for noisy data. IEEE Trans. Signal Process. **61**(22), 5620–5632 (2013)
28. N. Gillis, R.J. Plemmons, Sparse nonnegative matrix underapproximation and its application to hyperspectral image analysis. Linear Algebra Appl. **438**, 3991–4007 (2013)
29. Q. Zhang, H. Wang, R.J. Plemmons, V.P. Pauca, Tensor methods for hyperspectral data analysis: a space object material identification study. J. Opt. Soc. Am. A **25**(12), 3001–3012 (2008)
30. N. Gillis, R.J. Plemmons, Dimensionality reduction, classification, and spectral mixture analysis using non-negative underapproximation. Opt. Eng. **50**(2), 027001 (2011)
31. D.L. Donoho, V. Stodden, When does non-negative matrix factorization give a correct decomposition into parts?, in *Proceedings of the NIPS* (MIT Press, 2003)
32. S. Moussaoui, D. Brie, J. Idier, Non-negative source separation: range of admissible solutions and conditions for the uniqueness of the solution, in *Proceedings of IEEE International Conference on Acoustics, Speech, and Signal Processing, ICASSP*, vol. 5 (2005), pp. 289–292
33. N. Gillis, Sparse and unique nonnegative matrix factorization. J. Mach. Learn. Res. **13**, 3349–3386 (2012)
34. K. Huang, N.D. Sidiropoulos, A. Swami, Non-negative matrix factorization revisited: uniqueness and algorithm for symmetric decomposition. IEEE Trans. Signal Process. **62**(1), 211–224 (2014)
35. G. Casalino, N.D. Buono, C. Mencar, Subtractive clustering for seeding non-negative matrix factorizations. Inf. Sci. **257**, 369–387 (2014)
36. I. Meganem, Y. Deville, S. Hosseini, H. Carfantan, M. Karoui, Extraction of stellar spectra from dense fields in hyperspectral MUSE data cubes using non-negative matrix factorization, in *Proceedings of IEEE international Workshop WHISPERS, Lisbon, Portugal* (2011)
37. I. Meganem, S. Hosseini, Y. Deville, Positivity-based separation of stellar spectra using a parametric mixing model, in *Proceedings of the 21th European Signal Processing Conference (EUSIPCO), Marrakech, Morocco* (2013)

Chapter 7
NMF in MR Spectroscopy

T. Laudadio, A.R. Croitor Sava, Y. Li, N. Sauwen, D.M. Sima and S. Van Huffel

Abstract Nowadays, magnetic resonance spectroscopy (MRS) represents a powerful nuclear magnetic resonance (NMR) technique in oncology since it provides information on the biochemical profile of tissues, thereby allowing clinicians and radiologists to identify in a non-invasive way the different tissue types characterising the sample under investigation. The main purpose of the present chapter is to provide a review of the most recent and significant applications of non-negative matrix factorisation (NMF) to MRS data in the field of tissue typing methods for tumour diagnosis. Specifically, NMF-based methods for the recovery of constituent spectra in ex vivo and in vivo brain MRS data, for brain tissue pattern differentiation using magnetic resonance spectroscopic imaging (MRSI) data and for automatic detection and visualisation of prostate tumours, will be described. Furthermore, since several NMF implementations are available in the literature, a comparison in terms of pattern detection accuracy of some NMF algorithms will be reported and discussed, and the NMF performance for MRS data analysis will be compared with that of other blind source separation (BSS) techniques.

Keywords NMF · MRS · MRSI · Nosologic imaging · Brain tumours · Prostate cancer

---

T. Laudadio (✉)
Istituto per le Applicazioni del Calcolo "M. Picone" (IAC), Consiglio Nazionale delle Ricerche, Bari, Italy
e-mail: t.laudadio@ba.iac.cnr.it

A.R. Croitor Sava · N. Sauwen · D.M. Sima · S. Van Huffel
Department of Electrical Engineering (ESAT), STADIUS Center for Dynamical Systems, Signal Processing and Data Analytics, KU Leuven, Leuven, Belgium

A.R. Croitor Sava · N. Sauwen · D.M. Sima · S. Van Huffel
iMinds Medical Information Technologies, Leuven, Belgium

Y. Li
School of Electronic Engineering, University of Electronic Science and Technology of China, Chengdu, China

G.R. Naik (ed.), *Non-negative Matrix Factorization Techniques*,
Signals and Communication Technology, DOI 10.1007/978-3-662-48331-2_7

7.1 Introduction

NMR is widely used in oncology as a non-invasive diagnostic tool to detect the presence of tumour regions in the human body. One application of NMR is magnetic resonance imaging (MRI), which is applied in routine clinical practice to localise tumours and determine their size. MRI is able to provide an initial diagnosis, but its ability to delineate anatomical and pathological information is significantly improved by its combination with another NMR application, known as MRS. The latter provides information on the biochemical profile of tissues by exhibiting peaks at specific frequencies representing the molecular composition of the tissue under investigation, which appear in the MR signal as a spectrum of resonances. MRS can thus provide relevant information on cancer metabolism, allowing clinicians and radiologists to identify the different tissue types characterising the considered sample, and to study the biochemical changes underlying a pathological condition [1–3]. The simultaneous acquisition of several MR signals to produce spatially localised spectra within the organ of interest is referred to as MRSI. The analysis of MRS(I) data could represent a powerful diagnostic tool as it could be used to detect and localise the presence of a tumour, to classify its nature and to study the tumour metabolism in order to evaluate the response to possible therapies [4, 5]. Nevertheless, the clinical application of MRS(I) is still difficult because the analysis of MRS(I) data requires a lot of expertise as well as time-demanding, user-dependent and sophisticated processing procedures aimed at improving the quality of the acquired data and at extracting the information of interest. Therefore, automatic, accurate and efficient processing algorithms are needed in order to introduce MRS(I) in daily clinical practice.

Recently, NMF [6] has successfully been applied to brain as well as to prostate MRS(I) data in order to extract characteristic metabolic tissue patterns and to describe their contribution (abundance) to the profile of each spectrum. In particular, the studies described in [7–14] show that the application of NMF allows to overcome some of the main limitations that still affect most tissue segmentation and classification techniques of MRS(I) data available in the literature, i.e., need of prior knowledge, lack of automation and inability to properly describe tissue mixtures since each tissue volume element (or, equivalently, voxel), where the MRS signal is measured, is only labelled as being tumourous or non-tumorous.

On the opposite, extensive simulation and in vivo studies show that the proposed NMF-based methods are able to automatically detect accurate patterns for tumour, healthy and other interesting tissues without the need of any prior tissue model. Furthermore, they do not require massive MRS(I) data pre-processing aimed at removing artefacts, instrumental limitations, noise, etc., since NMF is able to detect the presence of uninteresting components as separate sources [13, 14].

The main purpose of the present chapter is to provide a review of the most recent and significant applications of NMF to MRS(I) data in the field of tissue typing methods for tumour diagnosis. Specifically, NMF-based methods for the recovery of constituent spectra in ex vivo and in vivo brain MRS data [7, 8], for brain tissue pattern differentiation using MRSI data [9–12], and for automatic detection and visualisation of prostate tumours [13, 14], will be described. Furthermore, since several NMF implementations are available in the literature, a comparison in terms of pattern detection accuracy of some NMF algorithms will be reported and discussed, and the NMF performance for MRS data analysis will be compared with that of other BSS techniques.

7.2 NMF: Basic Principles and Its Application to Ex Vivo and In Vivo Brain MRS Data for Tissue Pattern Extraction

NMF is a BSS technique imposing non-negative constraints on the extracted sources and corresponding weights. Specifically, given a non-negative matrix $X \cdot R^{m \times n}$, and an integer k $(0 < k < min(m, n))$, two non-negative matrices $W \cdot R^{m \times k}$ and $H \cdot R^{k \times n}$, minimising the functional

$$f(W, H) = \frac{1}{2} \|X - WH\|_F^2 \tag{7.1}$$

are estimated, where the subscript F stands for Frobenius norm [15]. It is worth noticing that, given the non-convex nature of $f(W, H)$, its minimisation may lead to the estimation of local minima and does not admit unique solution [16]. In order to deal with such a problem, a variety of NMF algorithms have been proposed in the literature and comparative studies of such implementations in terms of performance are often reported in theoretical as well as applicative studies [17]. Also in this chapter several NMF implementations will be considered and their performance will be compared in terms of source detection accuracy.

7.2.1 Application of NMF to Brain HR-MAS Data

In [7], NMF was applied to ex vivo high-resolution magic angle spinning (HR-MAS) spectra in order to extract characteristic tissue patterns of brain glial tumours and to quantify their contribution to each of the given spectra. HR-MAS represents an additional NMR spectroscopy technique that allows the identification of a higher number of important metabolites compared to standard MRS. For this reason, HR-MAS spectra are frequently used for metabolic analysis of tissue samples and, in particular, they have recently been used for brain tissue characterisation and tumour diagnosis [18], thereby representing a valuable alternative to histopathology which is time and effort expensive.

In general, brain glial tumours are heterogeneous and, for the high grade cases, they are also infiltrative. Therefore, HR-MAS spectra profiles belonging to the same brain tumour tissue type could show a significant variability [19]. Specifically, glial tumours might present contribution from the following different tissues: necrosis, highly cellular cells, healthy tissue and border tissue that contains tumour and apparently healthy brain cells. The observed spectra can then be considered a combination of different constituent sub-spectra, also referred to as sources, and the overall contribution of a tissue type to a spectrum is proportional to its abundance, i.e., to its proportion in the entire mixed tissue sample. As suggested in [7], the estimation of the aforementioned HR-MAS sub-spectra and corresponding abundances could be formulated as a BSS problem. More precisely, given n HR-MAS spectra, each consisting of m data points, they could be stacked as column vectors in an $m \times n$ matrix X. Then, the observed spectra can be described as follows:

$$X = WH + N \tag{7.2}$$

where the columns in W represent the unknown constituent tissue sources and the rows in H represent the corresponding concentrations or abundances. N represents additive noise. Given the non-negative nature of the HR-MAS data in [7], consisting of magnitude spectra, and the possible correlation that could characterise the profiles of the constituent tissues, two different NMF implementations were considered: an NMF algorithm exploiting sparsity constraints (NMFSC) [20], and one based on a convex analysis of mixtures of non-negative sources (CAMNS) [21]. Furthermore, the results obtained by NMFSC and CAMNS were compared with those ones provided by non-negative independent component analysis (nICA) [22], an additional non-negative BSS technique that imposes sources to be non-negative as well as mutually independent.

The aforementioned BSS algorithms were applied to HR-MAS spectra of brain tissue samples from 52 patients affected by a glial tumour with different degrees of malignancy. Such data were provided by the database of an European project entitled eTUMOUR (URL: http://www.etumour.net/). More precisely, the above algorithms were applied to the full HR-MAS spectra (consisting of 716 data points) as well as to metabolic feature vectors of different lengths ($n = 19$ and $n = 8$) extracted from the above spectra by peak integration [23] in order to study the influence of the input space (X matrix) on the algorithm performance. Specifically, the feature vector of length $n = 19$ contained all the estimated metabolite concentrations, and the one of lengths $n = 8$ was obtained by the previous vector by selecting only the concentrations of the most representative metabolites as reported in [19, 24].

To evaluate the source detection accuracy of NNMFSC, CAMNS and nICA, the correlation coefficient between the results provided by the above BSS techniques and reference tissue models provided by histopathology were computed. In particular, the correlation coefficients obtained by setting $k = 3$ (i.e., number of assumed pure tissues) for the glioblastoma (GBM) group when considering the different input spaces are displayed in Table 7.1. Such values show that NNMFSC and CAMNS provide a very accurate necrosis tissue source, independently of the input space dimension. A similar value is obtained by nICA as well when applied to the input space consisting of feature vectors of length $n = 8$. Concerning the detection of the tumour tissue source, all methods are less accurate than in the previous case and all

Table 7.1 Correlation coefficients between the obtained tissue sources and the reference tissue models for different input feature spaces [7]

	$n = 716$	$n = 19$	$n = 8$
NNMFSC			
Necrotic	0.97	0.97	0.99
Tumour	0.69	0.68	0.72
Border	0.91	0.92	0.92
CAMNS			
Necrotic	0.98	0.99	0.97
Tumour	0.62	0.61	0.71
Border	0.89	0.54	0.88
nICA			
Necrotic	0.81	0.93	0.98
Tumour	0.68	0.67	0.73
Border	0.83	0.51	0.68

of them perform best when applied to the aforementioned input space. Finally, the best border tissue source is provided by NNMFSC. Figure 7.1 shows the three tissue sources obtained by NNMFSC when applied to the magnitude spectra (left column) along with the corresponding reference tissue models provided by histopathology (right column). Looking at the overall performance of the considered methods, the

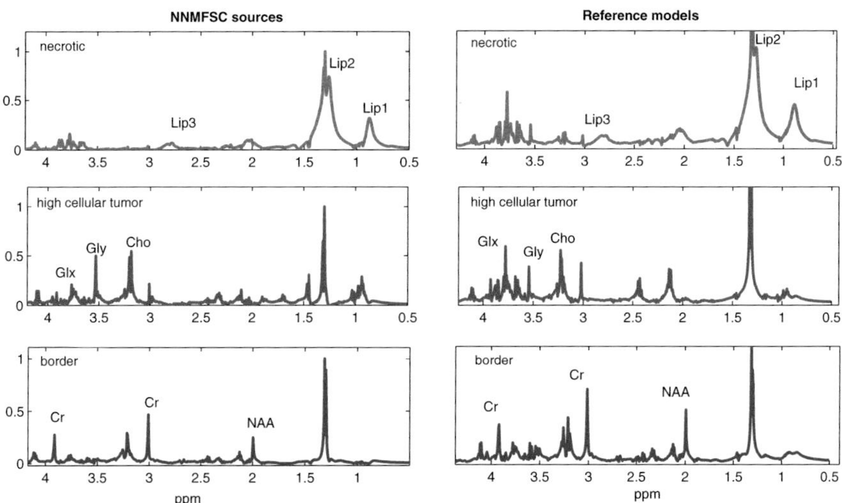

Fig. 7.1 *Left column* tissue sources obtained by NNMFSC when applied to magnitude spectra. *Right column* reference tissue models provided by histopathology. Lip1, Lip2 and Lip3 stand for lipids, Glx for glutamine + glutamate, Gly for glycine, Cho for choline and Cr for creatine [7]

results show that the sources provided by NNMFSC are the closest ones to the reference tissue models. CAMNS performs quite similarly to NNMFSC, probably because the two methods rely on the same principles, i.e., they impose sparsity and do not require sources to be mutually independent or uncorrelated. Still, with CAMNS, the assumption of sparsity is too strong since when a source, such as border and highly active tumour tissues, exhibits a non-zero value for a feature, all the other sources must have a zero value. On the other hand, this assumption is appropriate when considering the necrotic tissue source, which is characterised by high levels of lipids, while the other metabolic components are mostly embedded in noise and, therefore, are close to zero. Finally, nICA imposes the sources to be mutually independent and, consequently, uncorrelated. Such an assumption is too strong and not suitable for the considered data analysis since, as already described above, HR-MAS spectra from different tissue types can be correlated.

7.2.2 Application of NMF to Brain MRS Data

Recently, in [8] NMF has also been applied to single voxel brain MRS data measured in patients affected by brain tumour. The studies were carried out on data provided by an international multi-centre database [25] developed in the context of the European research project entitled INTERPRET [26]. In [8], the performance of a variety of NMF methods was investigated and compared when applied to the aforementioned MRS data to detect the main constituent sub-spectra representing characteristic pure tissue types in brain tumours. Furthermore, the NMF method characterised by the best performance in terms of tissue pattern detection was afterwards exploited, as pre-processing step, for dimensionality reduction in classical supervised classification algorithms commonly applied to MRS data.

The above studies show that the best performance was obtained by an NMF implementation known as CONVEX-NMF [27]. This algorithm, differently from the conventional ones available in the literature, imposes non-negative constraints only on the mixing matrix H, while entries in X and W are allowed to assume positive as well as negative values. Such a property makes CONVEX-NMF suitable for processing data of mixed sign. Indeed, if the given MRS data contain negative peaks, as those ones considered in [8], the detected sources will contain negative peaks as well, thereby avoiding the loss of important biochemical information which normally occurs when making the data non-negative prior to the application of conventional NMF methods. For instance, when processing long-echo time spectra, CONVEX-NMF allows the detection of some important metabolites, such as lactate, alanine and glutamine + glutamate, which play a crucial role in discriminating the different brain tumour types. Such a discrimination would not be possible by applying NMF algorithms imposing non-negative constraints on H as well as on X and W, since they usually truncate negative values occurring in the estimated sources to zero, with the consequent loss of information related to relevant negative peaks.

Concerning the application of CONVEX-NMF as pre-processing step for dimensionality reduction in supervised classification, the studies described in [8] show that it provides, in a fully unsupervised way, results comparable to and, in some cases, more accurate than those ones obtained by applying standard supervised classification approaches reported in the literature, such as for instance the one based on combining principal component analysis [28] with linear discriminant analysis [29].

7.3 NMF Applied to Brain MRSI Data for Tissue Pattern Differentiation

In [9], CONVEX-NMF has also been applied to brain MRSI data measured in tumour-bearing mice in order to extract sources describing characteristic pure tissues and, after determining the tumour tissue pattern, to delimit and visualise the tumour area in a fully unsupervised mode. As in [8], the performance of CONVEX-NMF has been compared with that of other NMF implementations, and the results of such studies confirm that CONVEX-NMF performs best in terms of source detection, thereby providing an accurate tumour region delimitation. The method proposed in [9] represents a significant achievement in tumour detection and visualisation since, differently from other approaches available in the literature, it is unsupervised and, therefore, does not require any label for tumour or normal tissue to find the desired sources. Nevertheless, it only allows the separation of tumour versus non-tumour tissue, without taking into account that the detection of other characteristic pure tissues is of crucial importance for therapy planning. Furthermore, up to our knowledge, the method has not been tested on MRSI data measured in human brain.

Alternative applications of NMF to human brain MRSI data have been proposed in [10–12]. More precisely, in these studies NMF has been embedded into a hierarchical scheme aimed at extracting the constituent sub-spectra from the given data and to visualise their distribution. It is worth highlighting that the hierarchical approaches proposed in [10–12] show significant differences. First of all, the studies described in [10] were carried out on long-echo time signals, while short-echo time signals were considered in [11, 12]. Second, they made use of different NMF algorithms. Specifically, in [10] the authors proposed a constrained version of NMF which forces negative amplitude values, occurring in the recovered sources and abundance distributions, to be approximately equal to zero. In [11, 12], the hierarchical alternating least squares (HALS) algorithm [30–32] was considered since in a preliminary study it was proven to be superior to other NMF implementations when applied to simulated brain MRSI data [33]. Last but not the least, the main purpose of the algorithm proposed in [10] consisted in separating abnormal brain tissue from the normal one. However, it is well known that for GBMs there exist two distinct patterns for abnormal tissue, representing tumour and necrosis tissue. In [11], a novel hierarchical non-negative matrix factorisation (hNMF) method was developed in order to accurately estimate the three most important tissue patterns for GBMs (normal, tumour and necrosis). For shortness' sake, below only the latter algorithm is described.

7.3.1 hNMF for GBM

Let us here assume that the model describing the data is (7.2). The algorithm hNMF consists in the following three main steps:

Step 1. First-level NMF: NMF is applied to the data matrix X by setting $k = 2$. Two spectral profiles and their corresponding h-maps (H_{normal} and $H_{abnormal}$) are recovered. The spectral sources obtained are automatically assigned to 'normal' and 'abnormal' tissue (W_{normal} and $W_{abnormal}$) based on the ratio between the metabolites N-acetylaspartate (NAA) and lipids (Lips) (where the values of NAA and Lips are estimated as the maximum intensity in the frequency regions around 2.01 and 1.3 ppm, respectively). The source with the higher NAA/Lips ratio corresponds to normal tissue and the source with the lower NAA/Lips ratio corresponds to abnormal tissue [34].

Step 2. Second-level NMF: NMF is performed with two sources repeatedly, on several sets of voxels, and the best result is chosen, as follows. Let us consider a variable threshold t on the magnitude of the $H_{abnormal}$ map, according to which a mask $f(t)$ is created, i.e. a selection of 'abnormal' voxels having values of $H_{abnormal}$ larger than t. For any fixed value of t, NMF with $k = 2$ is applied to the set of currently selected 'abnormal' spectra, leading to two sources, $W^1_{abnormal}(t)$ and $W^2_{abnormal}(t)$. In order to choose the best mask $f(\hat{t})$ from a set of gradually decreasing masks, a reasonable trade-off is sought, such that W_{normal}, $W^1_{abnormal}(\hat{t})$ and $W^2_{abnormal}(\hat{t})$ are mutually least correlated, i.e.

$$\hat{t} = \arg\min(corr1(t) + corr2(t) + corr3(t)) \tag{7.3}$$

where

$$\begin{aligned} corr1(t) &= corr(W_{normal}, W^1_{abnormal}(t)) \\ corr2(t) &= corr(W_{normal}, W^2_{abnormal}(t)) \\ corr3(t) &= corr(W^1_{abnormal}(t), W^2_{abnormal}(t)) \end{aligned} \tag{7.4}$$

and $corr(x, y) = \frac{(x-\bar{x})^T(y-\bar{y})}{\|x-\bar{x}\|\cdot\|y-\bar{y}\|}$ for any two vectors x and y of the same length, while $\bar{x}$ and $\bar{y}$ are the means of x and y, respectively.

The profiles $W^1_{abnormal}(\hat{t})$ and $W^2_{abnormal}(\hat{t})$ are automatically assigned to necrosis and tumour tissue ($W_{necrosis}$ and W_{tumour}) based on the choline (Cho)-to-Lips ratio (where such values are estimated as the maximum intensity in the frequency regions around 3.22 and 1.3 ppm, respectively). Based on previous studies [34], the source with the smaller Cho/Lips ratio is assigned to necrosis and the other is assigned to tumour.

Step 3. Non-negative least squares (NNLS) re-estimation: NNLS [35] is applied to the grid considered in Step 1 using the sources W_{normal}, W_{tumour} and $W_{necrosis}$ to re-estimate the corresponding h-maps.

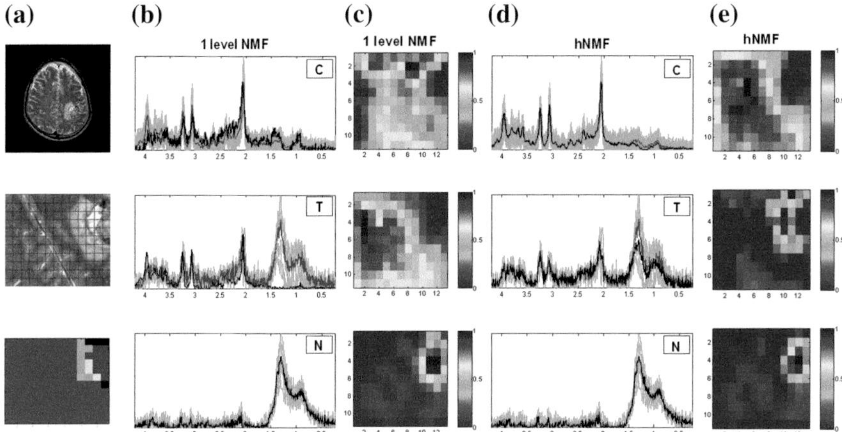

Fig. 7.2 Tissue pattern differentiation on 1H MRSI data from a patient with GBM: C, T and N represent normal, tumour and necrosis, respectively. **a** *First row* anatomical T2-weighted MR image. Clear tumour and necrosis regions are visible in the image. *Second row* selected voxels of interest. *Third row* expert labelling, where *red* indicates N, *yellow* indicates N/T, *green* indicates T, *cyan* indicates C/T, *blue* indicates C and *black* indicates spectra of poor quality. **b, c** Results of one-level NMF. **b** The recovered sources are shown in *black*. Spectra in *red* are the 'ideal' spectra according to expert labelling. Spectra in *green* show all the tissue-specific spectra labelled by the experts. **c** Corresponding h-maps for each spectral profile given in (**b**). **d, e** Results of hNMF. **d** The recovered sources for C, T and N are shown in *black*, overlaid over the 'ideal' spectra in *red*. Spectra in *green* show all the tissue-specific spectra labelled by the experts. **e** Corresponding h-maps which can be compared with the anatomical image and expert labelling [11]

The performance of the method was tested on simulated as well as on in vivo short-echo MRSI GBM data because GBM represents a very heterogeneous and aggressive brain tumour. Furthermore, it was compared with that of one-level NMF applied with $k = 3$, i.e. number of pure tissue patterns to be detected, and with expert labelling for the in vivo tests. As shown in Fig. 7.2, where the results obtained for an in vivo example are reported, hNMF significantly outperforms the one-level NMF approach in terms of tissue pattern detection and visualisation of the corresponding areas. The reason why the one-level NMF fails is probably due to the fact that the tissue-specific spectral profile of tumour is not sufficiently uncorrelated from a linear combination of the other patterns representing normal and necrosis tissue. On the opposite, hNMF looks for the three most distinct (but possibly correlated) tissue-specific spectral profiles by optimising a mask on the abnormal region, which is meant to exclude voxels containing a mixture of tumour and normal tissue patterns. The mixing coefficients are finally computed as the h-maps corresponding to each tissue pattern.

As already mentioned above, compared to the previous studies described in [10], hNMF presents several novelties: first of all, it was applied to short-echo time spectra that are able to provide more diagnostic information than long-echo time spectra, but are potentially more challenging because of the more complex spectral profiles;

hNMF was applied directly to brain tissue data and was able to differentiate three tissue patterns instead of only two (i.e. normal and abnormal) in the region encompassing the tumour; in the hierarchical application of NMF described in [10], only a fixed threshold was used to generate a mask for data selection, while in [11] an adaptive threshold selection was carried out; the NNLS re-estimation, added as last step in hNMF, avoids the loss of information at the border between tissue patterns, which may otherwise occur as a result of thresholding. Finally, the potential of hNMF is not limited to tissue pattern differentiation of GBMs. In principle, it could be adapted to solve other problems requiring more than two sources or more levels.

Recently, as described in [12], an additional step has been introduced into hNMF aimed at providing unsupervised nosologic images, where all the tissues of interest are simultaneously visualised by means of different colours. More precisely, each h-map provided by Step 3 (normalised between 0 and 1) is encoded as a colour channel into an RGB image, thereby obtaining a single image where red, green and blue, respectively, represent necrosis, tumour and normal tissue, while their possible mixture represents mixed tissues. The mixed colours of nosologic images provide a way to interpret tumour infiltration.

7.4 NMF Applied to Prostate MRSI Data for Automatic Detection and Visualisation of Tumours

In [13, 14], preliminary studies have been carried out on the application of NMF to simulate as well as to in vivo three-dimensional 3 T long-echo MRSI prostate data in order to extract characteristic patterns for tumour and benign tissue and to visualise their spatial distribution. In particular, in [14], a hierarchical approach is proposed which differs from those ones applied to brain data in [10–12]. Indeed, since the detection of only two tissue patterns is generally required for prostate data, i.e., tumour and benign tissue, NMF could be directly applied to the MRSI data set by setting $k = 2$. Nevertheless, as shown by extensive simulation studies, performing only one NMF level with $k = 2$ does not provide reliable results. Higher quality tissue patterns are obtained by embedding NMF into the hierarchical scheme (HNMF) as described below:

- The MRSI data are arranged into a matrix X, containing spectra as columns, and NMF is applied to X by setting $k = 2$. Two source signals and their corresponding weights are obtained.
- The given MRSI voxels are divided into two new data sets by assigning each voxel to the source signal characterised by the maximum weight in that voxel.
- The procedure is repeated after building new matrices X from spectra pertaining to each of the subsets obtained above.

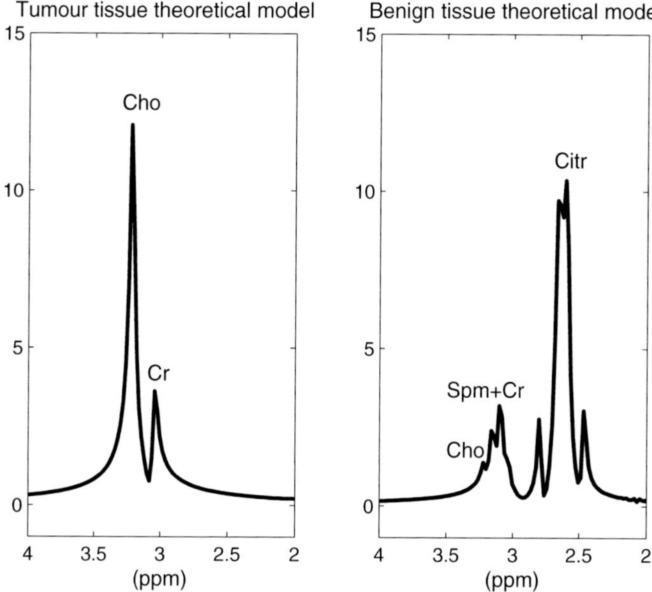

Fig. 7.3 Characteristic prostate magnitude spectra assumed for aggressive tumour tissue (*left*) and benign tissue (*right*). Cho stands for choline, Spm for spermine, Cr for creatine and Citr for citrate. The spectrum representing aggressive tumour tissue exhibits a high contribution of choline and no citrate, while benign tissue is represented by a spectrum characterised by a high citrate contribution, a small choline peak and contribution by the spermine model signal in the interval (3–3.2 ppm) [14]

In [14], several NMF levels, corresponding to increasingly finer partitions of the initial MRSI data, are applied, thereby obtaining a set of signal sources among which the two most significant ones, representing tumour and benign tissues, can be selected based on the prior knowledge that the choline-to-citrate ratio should be high for the source representing tumour tissue and low for the source representing benign tissue (see Fig. 7.3 where characteristic prostate magnitude spectra for aggressive tumour tissue (left) and benign tissue (right) are shown) [36].

As in [12], the selected tissue patterns can afterwards be encoded as different colour channels into an RGB image, and their spatial distribution can be visualised by means of different colours into a single image. As example, Fig. 7.4 shows a simulated dataset, consisting of a 10×10 grid of voxels (left picture), mimicking a prostate tumour. The contribution of the aggressive tumour model in each voxel of the above grid is displayed in the same figure (right picture). More precisely, the contribution of the tumour model to each spectrum, or voxel, in the grid can be retrieved from the colour bar displayed on the right-hand side of the image by looking for the value corresponding to the colour intensity present in the considered voxel.

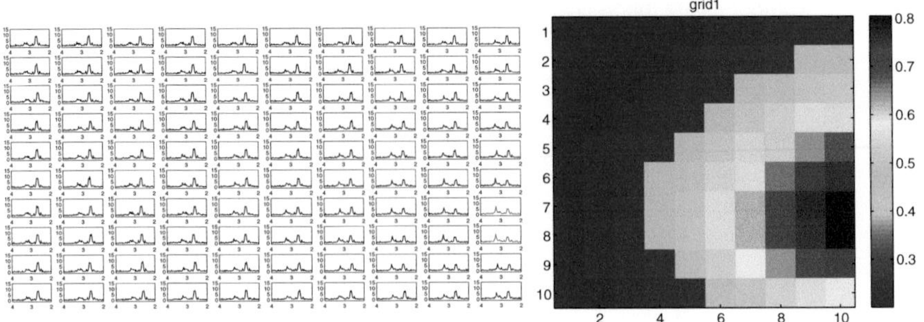

Fig. 7.4 *Left* simulated dataset. The most aggressive tumour spectra are displayed in *red* and they exhibit a low citrate contribution. *Right* tumour source spreading [14]

In Fig. 7.5, we first show the tumour and benign tissue patterns obtained by performing only one NMF level with $k = 2$ (left plots) and the final patterns provided by the hierarchical approach, i.e. by HNMF (right plots), for one simulation run when considering a medium noise level. The NMF implementation available in Matlab (a registered trademark of The MathWorks Inc., Natick, MA, USA), which is based on an alternating least squares scheme [17], and then here denoted by ALS, was applied.

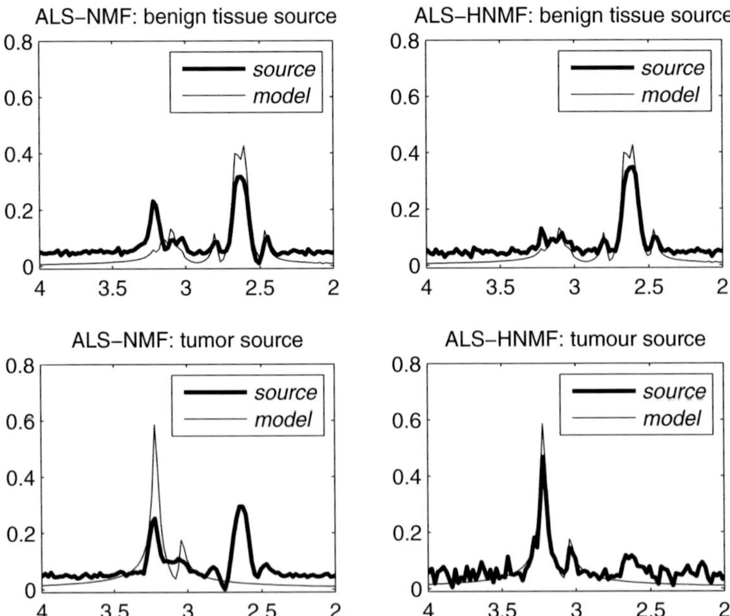

Fig. 7.5 Sources obtained by applying one level of ALS-NMF with $k = 2$ (*left*) and final tissue patterns provided by ALS-HNMF (*right*) superimposed to the theoretical tumour and benign tissue models. The X axis represents frequency values in ppm [14]

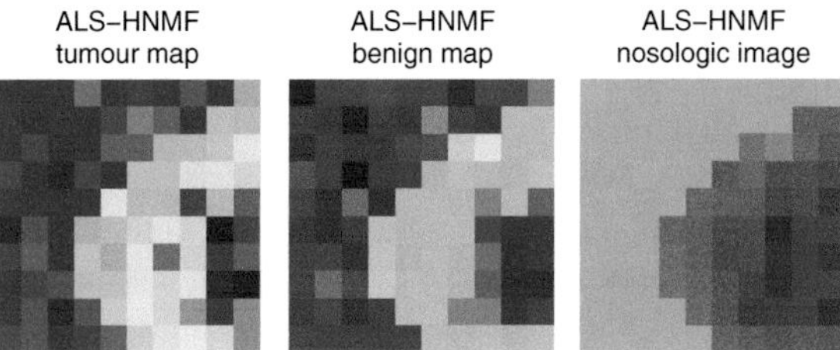

Fig. 7.6 Tissue maps and final nosologic images obtained by NNLS by exploiting the final patterns provided by ALS-HNMF. In the nosologic images, the pathological and benign regions are visualised in *red* and *green*, respectively [14]

In the plots, the patterns are superimposed to the theoretical tumour and benign tissue models. Figure 7.5 clearly shows that the patterns provided by one NMF level (i.e. by ALS-NMF) are very close to each other in shape and both of them represent more mixed tissue than tumour or benign tissue. On the contrary, the tissue patterns obtained by applying the hierarchical approach (ALS-HNMF) are of higher quality as they are very close to the theoretical models represented by the spectra in Fig. 7.3.

Figure 7.6 shows the tissue maps and corresponding nosologic images obtained by applying NNLS between the final tissue patterns obtained by ALS-HNMF and the data matrix X. In the nosologic image (right image), the pathological region is visualised in red and it matches well to that one displayed in the right image of Fig. 7.4, confirming the high degree of accuracy of the proposed NMF-based tissue characterisation approach.

7.5 Conclusions and Future Directions

In this chapter, several NMF-based methods representing the state-of-art in the field of tissue pattern detection and classification of brain and prostate MRS(I) data have been outlined. In particular, algorithms for the recovery of constituent spectra in ex vivo and in vivo brain MRS data, for brain tissue pattern differentiation and for automatic detection and visualisation of prostate tumours using MRSI data, have been described.

All these methods show that NMF is able to extract accurate characteristic tissue patterns from the considered data without the need of any prior tissue model and without imposing strong statistical constraints, such as mutual independence and/or uncorrelation among sources. Furthermore, when considering MRSI data, a fast and automatic visualisation of the pathological area is possible by applying NNLS between the final tissue patterns obtained by NMF and the given data matrix. This approach allows to overcome some of the main limitations that still affect most tissue segmentation and classification techniques of MRS(I) data available in the literature. Indeed, nowadays, the analysis of such data is mainly aimed at building the so-called metabolite maps, and images in which the concentrations of the most relevant metabolites are displayed. Unfortunately, the estimation of metabolite concentrations, known as MRS(I) quantification [37–39], is still difficult as MRS(I) spectra are generally affected by noise, artefacts and overlapping peaks, and several time-demanding and user-dependent pre-processing steps are needed before any accurate quantification method could be applied. On the contrary, NMF is very efficient as it requires computational times of the order of a few seconds to extract the desired tissue patterns, and, furthermore, it does not require massive MRS(I) data pre-processing aimed at removing uninteresting spectral components since these are detected as separate sources. Nevertheless, some open problems still characterise the application of NMF to MRS(I) data and their solution represents a difficult but crucial task in order to introduce MRS(I) in routine clinical practice. For instance, as described in Sect. 7.2, the minimisation of (7.1) does not admit unique solution and, therefore, a multiplicity of NMF algorithms is proposed in the literature. Indeed, in the methods described in Sects. 7.3–7.5, different NMF implementations were applied to process the MRS(I) data under investigation and, in some studies, their performance was compared in terms of accuracy. The results show that different implementations could provide different results, but the conclusions of the above studies are only limited to the specific considered implementations and applications, while it would be interesting to carry out a comparative study aimed at assessing whether and which NMF algorithm performs best when applied to any kind of MRS(I) dataset.

A second issue consists in the choice of the optimal k value, representing the number of sources to extract when applying NMF. In most of the MRS(I) applications reported above, such a choice is based on a visual inspection of the sources obtained when applying the method with different k values. More precisely, starting from $k = 2$, k is gradually increased until representative tissue patterns are identified among the extracted sources and, at the same time, further increasing its value only provides redundant information. Although fast, the above approach requires a lot of expertise and user interaction. Currently, other selection criteria are under investigation aimed at providing the optimal value to assign to k automatically.

Finally, an important open problem is represented by the choice of the NMF levels to include in the hierarchical approaches proposed in [10, 14]. As for k, an automatic and, possibly, application-independent procedure would be desirable rather than the heuristic one currently adopted in the aforementioned studies [40].

Solving the above issues would make the application of NMF to MRS(I) data fully unsupervised and automatic, thereby representing an important achievement in the field of MRS(I) data tissue typing for non-invasive diagnosis of cancer.

Besides the aforementioned issues, a possible future research topic is represented by the inclusion of spatial information. Indeed, as described in the Introduction, MRSI provides biochemical information as well as spatial information about the sample under investigation and a natural assumption is that neighbour voxels contain similar tissue types. Previous studies carried out on other algorithms [41, 42] already showed that the simultaneous exploitation of the two types of information significantly improves the performance of the considered methods. Hence, it is reasonable to expect similar improvements by introducing spatial information in the NMF-based algorithms as well.

Another interesting topic could be represented by the application of NMF to multi-parametric MR data, where information are extracted by making use of more MR modalities (e.g. MRI, MRSI, perfusion-weighted imaging and diffusion-weighted imaging), and properly combined in terms of feature vectors. Also in this case, previous studies showed that exploiting multi-modal information yields more accurate segmentation and classification results [43]. Recently, an interesting unsupervised NMF-based approach has been proposed in [44], where MRI and MRSI data information are simultaneously exploited for detecting brain abnormalities. In conclusion, embedding spatial as well as multi-modal information into NMF would further improve the degree of reliability of the results and, therefore, would make the application of MRS(I) possible in routine clinical practice for diagnosis, monitoring and therapy planning purposes.

Acknowledgments The work of Teresa Laudadio is partly supported by the following projects: PRIN 2012 entitled SMaSIP (Structured matrices in Signal and Image Processing) and Progetto Premiale MIUR 2012 entitled MATHTECH.
The work of Yuqian Li is supported by Young Scientists Fund of the National Natural Science Foundation of China (Grant No. 61401068) and by the China Postdoctoral Science Foundation (Grant No. 2014M552341).
The work of Anca Croitor, Nicolas Sauwen, Diana Sima and Sabine Van Huffel is supported by Research Council KUL: GOA/10/09 MaNet, CoE PFV/10/002 (OPTEC); PhD/Postdoc grants. Flemish Government: FWO: projects: G.0427.10N (Integrated EEG-fMRI), G.0108.11 (Compressed Sensing) G.0869.12N (Tumor imaging) G.0A5513N (Deep brain stimulation); PhD/Postdoc grants; IWT: projects: TBM 080658-MRI (EEG-fMRI), TBM 110697-NeoGuard; PhD/Postdoc grants; iMinds Medical Information Technologies SBO 2014, ICON: NXT_Sleep; Flanders Care: Demonstratieproject Tele-Rehab III (2012–2014). Belgian Federal Science Policy Office: IUAP P7/19/(DYSCO, 'Dynamical systems, control and optimization', 2012–2017). Belgian Foreign Affairs-Development Cooperation: VLIR UOS programmes. EU: The research leading to these results has received funding from the European Research Council under the European Union's Seventh Framework Programme (FP7/2007–2013)/ERC Advanced Grant: BIOTENSORS (no. 339804). This paper reflects only the authors' views and the Union is not liable for any use that may be made of the contained information. Other EU funding: RECAP 209G within INTERREG IVB NWE programme, EU MC ITN TRANSACT 2012 (no. 316679), ERASMUS EQR: Community service engineer (no. 539642-LLP-1-2013).

References

1. S. Nelson, Multivoxel magnetic resonance spectroscopy of brain tumors. Mol. Cancer Ther. **2**(5), 497–507 (2003)
2. X. Leclerc, T. Huisman, A. Sorensen, The potential of proton magnetic resonance spectroscopy (1H-MRS) in the diagnosis and management of patients with brain tumors. Curr. Opin. Oncol. **14**, 292–298 (2002)
3. B. Pickett, J. Kurhanewicz, F. Coakley, K. Shinohara, B. Fein, M. Roach, Use of MRI and spectroscopy in evaluation of external beam radiotherapy for prostate cancer. Int. J. Radiat. Oncol. Biol. Phys. **60**(4), 1047–1055 (2004)
4. L. Kwock, J.K. Smith, M. Castillo, M.G. Ewend, F. Collichio, D.E. Morris, T.W. Bouldin, S. Cush, Clinical role of proton magnetic resonance spectroscopy in oncology: brain, breast, and prostate cancer. Lancet Oncol. **7**(10), 859–868 (2006)

5. S.J. Nelson, Magnetic resonance spectroscopic imaging. IEEE Eng. Med. Biol. **23**, 30–39 (2004)
6. D.D. Lee, H.S. Seung, Learning the parts of objects by non-negative matrix factorization. Nature **401**, 788–791 (1999)
7. A.R. Croitor Sava, C.M. Martinez-Bisbal, D.M. Sima, J. Calvar, V. Esteve, B. Celda, U. Himmelreich, S. Van Huffel, Quantifying brain tumor tissue abundance in HR-MAS spectra using non-negative blind source separation techniques. J. Chemom. **26**(7), 406–415 (2012)
8. S. Ortega-Martorell, P.J.G. Lisboa, Julià-Sapé M. Vellido, C. Arús, Nonnegative matrix factorisation methods for the spectral decomposition of MRS data from human brain tumours. BMC Bioinform. **13**, 38 (2012)
9. S. Ortega-Martorell, P.J. Lisboa, A. Vellido, R.V. Simões, M. Pumarola, M. Julià-Sapé, C. Arús, Convex non-negative matrix factorization for brain tumor delimitation from MRSI data. PLoS ONE **7**(10), e4 (2012)
10. P. Sajda, S. Du, T.R. Brown, R. Stoyanova, D.C. Shungu, X. Mao, L.C. Parra, Nonnegative matrix factorization for rapid recovery of constituent spectra in magnetic resonance chemical shift imaging of the brain. IEEE Trans. Med. Imaging **23**, 1453–1465 (2004)
11. Y. Li, D.M. Sima, S. Van Cauter, A.R. Croitor Sava, U. Himmelreich, Y. Pi, S. Van Huffel, Hierarchical non-negative matrix factorization (hNMF): a tissue pattern differentiation method for glioblastoma multiforme diagnosis using MRSI. NMR Biomed. **26**(3), 307–319 (2013)
12. Y. Li, D.M. Sima, S. Van Cauter, A.R. Croitor Sava, U. Himmelreich, Y. Pi, Y. Liu, S. Van Huffel, Unsupervised nosologic imaging for glioma diagnosis. IEEE Trans. Biomed. Eng. **60**(6), 1760–1763 (2013)
13. A.R. Croitor Sava, A. Wright, D.M. Sima, T. Laudadio, S. Van Huffel, A. Heerschap, U. Himmelreich, Automatic magnetic resonance spectroscopic imaging segmentation using blind source separation techniques. Lirias number: 421620, in *Proceedings ESMRMB 2013*, Toulouse, October 2013, pp. 330–331
14. T. Laudadio, A.R. Croitor Sava, D.M. Sima, A. Wright, A. Heerschap, S. Van Huffel, Hierarchical non-negative matrix factorization applied to in vivo 3T MRSI prostate data for automatic detection and visualization of tumours. Lirias number: 421618, in *Proceedings ESMRMB 2013*, Toulouse, October 2013, pp. 474–475
15. G.H. Golub, C.F. Van Loan, *Matrix Computations*, 4th edn. (The Johns Hopkins University Press, Baltimore, 2013)
16. H. Laurberg, M.G. Christensen, M.D. Plumbley, L.K. Hansen, S.H. Jensen, Theorems on positive data: on the uniqueness of NMF. Comput. Intell. Neurosci. 764206 (2008). doi:10.1155/2008/764206
17. M.W. Berry, M. Browne, A.N. Langville, V.P. Pauca, R.J. Plemmons, Algorithms and applications for approximate nonnegative matrix factorization. Comput. Stat. Data Anal. **52**, 155–173 (2007)
18. M.C. Martínez-Bisbal, L. Martí-Bonmatí, J. Piquer, A. Revert, P. Ferrer, J.L. Llácer, M. Piotto, O. Assemat, B. Celda, ^{1}H and ^{13}C HR-MAS spectroscopy of intact biopsy samples ex vivo and in vivo ^{1}H MRS study of human high grade gliomas. NMR Biomed. **17**(4), 191–205 (2004)
19. A.R. Croitor Sava, M.C. Martinez-Bisbal, S. Van Huffel, J.M. Cerda, D.M. Sima, B. Celda, Ex vivo high resolution magic angle spinning metabolic profiles describe intratumoral histopathological tissue properties in adult human gliomas. Magn. Reson. Med. **65**, 320–328 (2011)
20. H. Kim, H. Park, Sparse non-negative matrix factorizations via alternating non-negativity-constrained least squares for microarray data analysis. Bioinformatics **23**(12), 1495–1502 (2007)
21. H. Cha, W.K. Ma, C.Y. Chi, Y. Wang, A convex analysis framework for blind separation of non-negative sources. IEEE Trans. Signal Process. **56**(10), 5120–5134 (2008)
22. M.D. Plumbley, Algorithms for non-negative independent component analysis. IEEE Trans. Neural Netw. **14**(3), 534–543 (2003)
23. R. Meyer, M. Fisher, S. Nelson, T. Brown, Evaluation of manual methods for integration of in vivo phosphorus NMR spectra. NMR Biomed. **1**(3), 131–135 (1988)

24. L.L. Cheng, I.W. Chang, D.N. Louis, R.G. Gonzalez, Correlation of high-resolution magic angle spinning proton magnetic resonance spectroscopy with histopathology of intact human brain tumor specimens. Cancer Res. **58**(9), 1825–1832 (1998)
25. M. Julià-Sapé, D. Acosta, M. Mier, C. Arús, D. Watson, The interpret consortium: a multi-centre, web-accessible and quality control-checked database of in vivo MR spectra of brain tumour patients. Magn. Reson. Mater. Phys. **19**, 22–33 (2006)
26. A.R. Tate, J. Underwood, D.M. Acosta et al., Development of a decision support system for diagnosis and grading of brain tumours using in vivo magnetic resonance single voxel spectra. NMR Biomed. **19**(4), 411–434 (2006)
27. C. Ding, T. Li, M. Jordan, Convex and semi-nonnegative matrix factorizations. IEEE Trans. Pattern Anal. Mach. Intell. **99**(1), 5555 (2008)
28. I.T. Jolliffe, *Principal Component Analysis* (Springer, New York, 2002)
29. K.S. Opstad, C. Ladroue, B.A. Bell, J.R. Griffiths, F.A. Howe, Linear discriminant analysis of brain tumour (1)H MR spectra: a comparison of classification using whole spectra versus metabolite quantification. NMR Biomed. **20**(8), 763–770 (2007)
30. A. Cichocki, R. Zdunek, S. Amari, Hierarchical ALS algorithms for nonnegative matrix and 3D tensor factorization. Lect. Notes Comput. Sci. **4666**, 169–176 (2007)
31. A. Cichocki, A.H. Phan, Fast local algorithms for large scale nonegative matrix and tensor factorizations. IEICE Trans. Fundam. Electron. Commun. Comput. Sci. **3**, 708–721 (2009)
32. N. Gillis, Nonnegative matrix factorization complexity, algorithms and applications. Ph.D. thesis, Louvan-La-Neuve (2011)
33. Y. Li, D.M. Sima, S. Van Cauter, U. Himmelreich, Y. Pi, S. Van Huffel, Simulation study of tissue type differentiation using non-negative matrix factorization, in *Proceedings of BIOSIGNALS 2012, International Conference on Bioinspired Systems and Signal Processing*, Vilamoura, Algarve, February 2012, pp. 212–217
34. F.A. Howe, S.J. Barton, S.A. Cudlip, M. Stubbs, D.E. Saunders, J.R. Murphy, K.S. Opstad, V.L. Doyle, M.A. McLean, B.A. Bell, J.R. Griffiths, Metabolic profiles of human brain tumours using quantitative in vivo 1H magnetic resonance spectroscopy. Magn. Reson. Med. **49**, 223–232 (2003)
35. C.L. Lawson, *Solving Least-Squares Problems*, vol. 23 (Prentice-Hall, Englewood Cliffs, 1974), p. 161
36. K.M. Selnaes, I.S. Gribbestad, H. Bertilsson, A. Wright, A. Angelsen, A. Heerschap, M.B. Tessem, Spatially matched in vivo and ex vivo MR metabolic profiles of prostate cancer-investigation of a correlation with Gleason score. NMR Biomed. **26**(5), 600–606 (2013)
37. S.W. Provencher, Estimation of metabolite concentrations from localized in vivo proton NMR spectra. Magn. Reson. Med. **30**(6), 672–679 (1993)
38. J.B. Poullet, D.M. Sima, A.W. Simonetti, B. De Neuter, L. Vanhamme, P. Lemmerling, S. Van Huffel, An automated quantitation of short echo time MRS spectra in an open source software environment: AQSES. NMR Biomed. **20**(5), 493–504 (2007)
39. J.B. Poullet, D.M. Sima, S. Van Huffel, MRS signal quantitation: a review of time- and frequency-domain methods. J. Magn. Reson. **195**, 134–144 (2008)
40. N. Gillis, D. Kuang, H. Park, Hierarchical clustering of hyperspectral images using rank-two nonnegative matrix factorization (2013). arXiv preprint arXiv:1310.7441
41. T. Laudadio, P. Pels, L. De Lathauwer, P. Van Hecke, S. Van Huffel, Tissue segmentation and classification of MRSI data using canonical correlation analysis. Magn. Reson. Med. **54**, 1519–1529 (2005)
42. A.R. Croitor Sava, D.M. Sima, J.B. Poullet, A.J. Wright, A. Heerschap, S. Van Huffel, Exploiting spatial information to estimate metabolite levels in two-dimensional MRSI of heterogeneous brain lesions. NMR Biomed. **24**(7), 824–835 (2011)
43. M. De Vos, T. Laudadio, A.W. Simonetti, A. Heerschap, S. Van Huffel, Fast nosologic imaging of the brain. J. Magn. Reson. **184**, 292–301 (2006)
44. X. Liu, Y. Li, Y. Pi, S. Van Cauter, Y. Yao, J. Wang, A new algorithm for the fusion of MRSI & MRI on the brain tumour diagnosis, in *Proceedings ISMRM 2015*, Toronto, Ontario, 30 May–5 June 2015

Chapter 8
Time-Scale-Based Segmentation for Degraded PCG Signals Using NMF

F. Sattar, F. Jin, A. Moukadem, C. Brandt and A. Dieterlen

Abstract This article deals with the challenging problem of segmenting narrowly spaced cardiac events (S1 and S2) in noisy phonocardiogram (PCG) signals by using a novel application of NMF based on time-scale approach. A novel energy-based method is proposed for the segmentation of noisy PCG signals in order to detect cardiac events, which could be closely spaced and separated by noisy gaps. The method is based on time-scale transform as well as nonnegative matrix factorization (NMF) and the segmentation problem is formulated using the paradigm of binary statistical hypothesis testing. The energy of the Morlet wavelet transform and NMF output is employed as a test statistics for segmentation where the number of scales are selected based on the preferences calculated along the time-scales. The simulation results using real recorded noisy PCG data that provide promising performance with high overall accuracy on the segmentation of narrowly separated, high noisy signals by our proposed method.

Keywords Segmentation · PCG signals · Cardiac events · NMF · Wavelet transform

8.1 Introduction

The detection/classification of cardiac events using Phonocardiogram (PCG) signals (i.e., cardiac sounds) is vital for cardiac auscultation, which is the basis for heart examination. It provides important information about structural and functional

F. Sattar (✉)
Department of Electrical and Computer Engineering, University of Waterloo, Waterloo, ON, Canada
e-mail: fsattar@uwaterloo.ca

F. Jin
Department of Electrical and Computer Engineering, Ryerson University, Toronto, ON, Canada

A. Moukadem · C. Brandt · A. Dieterlen
MIPS Laboratory, University of Haute Alsace, 68093 Mulhouse Cedex, France

G.R. Naik (ed.), *Non-negative Matrix Factorization Techniques*,
Signals and Communication Technology, DOI 10.1007/978-3-662-48331-2_8

cardiac defects. One of the first steps in PCG analysis is to segment the heart sound signal into four parts: S1 (first heart sound), systole, S2 (second heart sound), and diastole. Identification of these two phases of the cardiac cycle and of the heart sounds with robust differentiation between S1 and S2 event in the presence of noise and/or additional heart sounds is the first challenging step toward automatic diagnosis of heart murmurs with the distinction of ejection and regurgitation murmur [1].

This cardiac sound of PCG signal are short, impulse-like events representing transitions between different hemodynamic phases of cardiac cycle [2]. The major signal components of PCG signal, S1 and S2, are located within the low frequency range, approximately between 10 and 750 Hz [3]. In the literature, adjacent S1 and S2 components of the PCG signals are distinguished from one another in terms of their time-domain [3] and frequency-domain characteristics [4]. Some recent advances in PCG segmentation include: the noise/spike segments detection of PCG signal using time-domain information and clustering approach [5], detection of cardiac abnormality through LS-SVM classifier [6], heart sound classification using S-transform-based neural network scheme [7], which all require accurate segmentation of heart sound events in PCG signals.

A major challenge of general sound segmentation lies in selecting robust acoustic features and learning models with these features in such a way that diverse sound types are differentiated. For temporal/spectrally overlapped signals, direct spectrum-based features are not adequate, due to its high dimensionality. Efficient information representation thus plays a critical role in understanding perception of sensory data as well as in audio segmentation. In the literature, linear generative model has been widely adopted for sound representation where the structure of the signals is modeled in terms of a linear superposition of basis functions. One way to find a reduced dimension feature using the linear generative model is to constrain both basis vectors and latent variables to be nonnegative so that non-subtractive combinations of basis vectors are used to model the observed data. The nonnegative matrix factorization (NMF) technique introduced by Lee and Seung [8], is a subspace method which finds a linear data representation with nonnegativity constraints. It attempts matrix factorization in a way that the factors have nonnegative elements by performing a simple multiplicative updating.

NMF is able to produce useful representations of real-world data and can, therefore, be applied here to solve the problem of PCG signals segmentation. Based on the observed PCG data, the NMF decomposes the data matrix into two basis matrices. This results in reduced representation of the original data where each feature is a linear combination of the original attribute set [9]. The NMF has low computational complexity and unlike time–frequency techniques, it is able to deal with both dense and sparse data sets [10, 11]. Some relevant applications of NMF include single channel source separation for drum transcription [12], and blind recovery of single channel mixed cardiac and respiratory sounds [13].

This article deals with the segmentation of signal components separated by narrow and noisy gap which has been represented by the segmentation of S1 and S2 components of the PCG signals. Given its nonstationary signal characteristics, continuous wavelet transform has been adopted here, to provide localized information for PCG segmentation in the time scale followed by using a discriminative NMF decomposition. The NMF decomposition makes this maximum-likelihood-based segmentation scheme robust against high noise and insensitive to the choice of threshold due to its discriminative nature. The proposed method thus aims to perform robust segmentation of the noisy PCG signals based on raw PCG recordings.

8.2 Preliminaries

8.2.1 Signal Model

Consider, a signal consists of L components, which may overlap in time or in frequency. We may express the signal, $s_o(k)$, as follows:

$$s_o(k) = \sum_{i=1}^{L} \alpha_i s_{oi}(k - k_i) \tag{8.1}$$

where $s_{oi}(k)$ is the ith component signal, α_i is its weight and k_i is its time location. The measured signal, $y_o(k)$, is corrupted by noise, $v_o(k)$.

$$y_o(k) = s_o(k) + v_o(k) \tag{8.2}$$

8.2.2 Basic Concept

If the waveforms do not overlap in time domain, then the segmentation of a L-component waveform involves determining the time location, k_i, and the duration, τ_i. If, on the other hand, the components do not overlap in the frequency domain, then determining the center frequency, ω_i, and the bandwidth, B_i, may segment the waveform. In general, to segment the waveform when there is temporal or spectral overlap, both the time and frequency domain parameters are computed. In other words, the signal components are segmented in the time–frequency plane.

In the time–frequency plot, a signal component may be represented by a rectangle with its side parallel to the time and the frequency axis. The center of the rectangle is located at (k_i, ω_i), its width is τ_i and its height is B_i. Two signal components are better resolved if the area of the rectangle is small and the two components may be segmented if there are two distinct peaks in the 3-D plot of time, frequency, and intensity. However, Heisenberg's inequality imposes a lower bound on the area of the rectangle

$$\tau_i B_i \geq \frac{1}{4\pi} \approx 0.08 \tag{8.3}$$

Only Gaussian function suitably dilated, translated, and modulated will achieve a minimum bandwidth-duration product.

The question arises as to whether there exists a transformation from time domain to time–frequency domain such that the components signals may efficiently be segmented by exploiting the time and the frequency domain properties, rather than merely time or frequency domain properties. For conceptual simplicity, and efficient implementation, we will restrict in the followings to a class of linear time–frequency transformation.

8.2.3 NMF

The idea of non-negative matrix factorization (NMF) [8] to decompose the data matrix, **R**, as

$$\mathbf{R} \approx \mathbf{DH} \tag{8.4}$$

where **D** and **H** refer to the nonnegative matrices which we call the dictionary (basis functions) and the code (coefficients), respectively. There exists different algorithms for computing NMF factorization [14–17]. Here, we use the method proposed in [15] due to its simple formulation and easy implementation. This NMF algorithm begins with randomly initialized matrices, **D** and **H**, and alternates the following updates until convergence

$$\mathbf{H} \leftarrow \mathbf{H} \bullet \left((\overline{\mathbf{D}}^T \mathbf{R}) ./ (\overline{\mathbf{D}}^T \overline{\mathbf{D}} \mathbf{H} + \lambda) \right) \tag{8.5}$$

$$\mathbf{D} \leftarrow \overline{\mathbf{D}} \bullet \left((\mathbf{R}\mathbf{H}^T + \overline{\mathbf{D}} \bullet (\mathbf{1}(\overline{\mathbf{D}}\mathbf{H}\mathbf{H}^T \bullet \overline{\mathbf{D}}))) ./ (\overline{\mathbf{D}}\mathbf{H}\mathbf{H}^T + \overline{\mathbf{D}} \bullet (\mathbf{1}\mathbf{R}\mathbf{H}^T \bullet \overline{\mathbf{D}}))) \right) \tag{8.6}$$

where $\overline{\mathbf{D}}$ is the columnwise normalized dictionary matrix, **1** is a square matrix with all elements equal to 1, '•' refers to pointwise multiplication, './' denotes the pointwise division, and λ parameter represents the degree of sparsity.

8.3 Proposed Method

The proposed segmentation method is based on time-scale characteristic of a linear transform, which may be expressed as a linear combination of some basis functions, $\{\phi(k, m)\}, k = 1, \ldots, N-1; m = 1, \ldots, M-1$, where k is the time index and m is the frequency index. Then using linear transform the time–frequency transformation of the signal $y_o(k)$ in (8.2) will take the following matrix form:

$$\mathbf{y}_k = \mathbf{W}_k^H \mathbf{y}_o(k) \tag{8.7}$$

where superscript H indicates conjugate transpose, $\mathbf{y}_k$ is $(M \times 1)$, $\mathbf{y}_o$ is $(N \times 1)$ and $\mathbf{W}_k$ is $(N \times M)$ matrices given by

$$\begin{aligned}\mathbf{y}_k &= [y(k,0)\ x(k,1) \cdots y(k,M-1)]^T \\ \mathbf{y}_o &= [y_o(k)\ y_o(k-1) \cdots y_o(k-N+1)]^T\end{aligned} \tag{8.8}$$

$$\mathbf{W}_k = \begin{bmatrix} \phi(k,0) & \cdots \phi(k,M-1) \\ \phi(k-1,0) & \cdots \phi(k-1,M-1) \\ \vdots & \cdots \vdots \\ \phi(k-N+1,0) & \cdots \phi(k-N+1,M-1) \end{bmatrix} \tag{8.9}$$

where $\bullet^T$ denotes transposition.

Now considering (8.1) and (8.2) and applying the above transformation yields

$$\mathbf{y}_k = \mathbf{s}_k + \mathbf{v}_k \tag{8.10}$$

where

$$\begin{aligned}\mathbf{s}_k = \mathbf{W}_k^H \mathbf{s}_o(k),\ \ \mathbf{v}_k &= \mathbf{W}_k^H \mathbf{v}_o(k) \\ \mathbf{s}_o(k) = [s_0(k)\ s_0(k-1)\ &\cdots s_0(k-N+1)]^T \\ \mathbf{v}_o(k) = [v_0(k)\ v_0(k-1)\ &\cdots v_0(k-N+1)]^T\end{aligned} \tag{8.11}$$

Note that the choice of the transform matrix, $\mathbf{W}_k$, or equivalently the basis functions, $\{\phi(k,m)\}$, is problem dependent, that is, it depends upon the requirement of the time, frequency, and time–frequency resolutions. In the sequel, Morlet wavelet transform which is given by [18].

$$\phi(k,m) = \pi^{-1/4} e^{j2\pi f_o k/m} e^{-k^2/2m^2} \tag{8.12}$$

is employed here because of its ability to provide different window lengths for signals composed of different frequencies, and the time-bandwidth product is minimal providing thereby an optimal time and frequency resolution.

8.3.1 Decision Rule

The segmentation of the signal components can be formulated as a binary hypothesis testing problem between two hypothesis, H_o and H_1 as

$$\begin{aligned}H_0 &: \mathbf{y}_k = \mathbf{v}_k \\ H_1 &: \mathbf{y}_k = \mathbf{s}_k + \mathbf{v}_k\end{aligned} \tag{8.13}$$

where $\mathbf{s}_k$, $\mathbf{v}_k$ and $\mathbf{y}_k$ are the transform coefficients (i.e., wavelet transform in our case) of the original signals, $s_o(k)$, $v_o(k)$, and $y_o(k)$, respectively, and $\mathbf{v}_k$ is zero-mean white Gaussian noise. At a given instant k, we have to decide if $\mathbf{y}_k$ contains a signal component $\mathbf{s}_k$, or $\mathbf{y}_k$ is only a noise $\mathbf{v}_k$.

We employ the maximum-likelihood approach, which provides generalized likelihood ratio test (GLRT) for the signal $\mathbf{y}_k$ and noise $\mathbf{v}_k$, in our case, with unknown covariances given by [19, 20].

$$l(\mathbf{y}_k) = ln\ p(\mathbf{y}_k, \hat{\mathbf{R}}_{yy}; H_1) - ln\ p(\mathbf{y}_k, \hat{\mathbf{R}}_{vv}; H_o) \begin{cases} > \gamma \text{ for } H_1 \\ \leq \gamma \text{ for } H_0 \end{cases} \tag{8.14}$$

where $\hat{\mathbf{R}}_{yy}$, $\hat{\mathbf{R}}_{vv}$ are the maximum likelihood estimates of the covariance under hypothesis H_1 and H_o. This GLRT test is implemented by a simple and efficient scheme based on NMF decomposition. For this the covariance matrix of $\mathbf{y}_k$, $\mathbf{R}_{yy}$, under the hypothesis H_1 is given by

$$\mathbf{R}_{yy} = E[\mathbf{y}_k \mathbf{y}_k^H] \tag{8.15}$$

where $E[\cdot]$ is the expectation operator.

Using NMF-decomposition of the measurement matrix $\mathbf{R}_{yy}$, we get

$$\tilde{\mathbf{R}}_{yy} \approx \mathbf{DH} \tag{8.16}$$

Using Singular-Value-Decomposition (SVD) of the matrix $\tilde{\mathbf{R}}_{yy}$, we get

$$\tilde{\mathbf{R}}_{yy} = \mathbf{UCU}^T, \quad \mathbf{C} = diag[\sigma_1^2, \ldots, \sigma_M^2] \tag{8.17}$$

where $\mathbf{U}$ is the eigenvector matrix of $\tilde{\mathbf{R}}_{yy}$ and $\sigma_1 \leq \sigma_2 \leq \cdots \leq \sigma_M$ are the corresponding singular values.

As in [19, 21] the log-likelihood function becomes

$$l(\mathbf{y}_k) = -\frac{1}{2}\sum_{i=1}^{M} ln\sigma_i^2 + \frac{M}{2} ln\sigma_i^2 + \frac{1}{2\sigma_v^2}\sum_{i=1}^{M}\left(\frac{\sigma_i^2 - \sigma_v^2}{\sigma_i^2}\right)|U_i \mathbf{y}_k|^2 \tag{8.18}$$

Combining the non-data dependent terms into the threshold and scaling, the test statistics $T(\mathbf{y}_k)$ for the GLRT becomes

$$T(\mathbf{y}_k) = \sum_{i=1}^{M} G_i |U_i \mathbf{y}_k|^2 \text{ where } G_i = \frac{\sigma_{si}^2}{\sigma_i^2}, \quad \sigma_{si}^2 = \sigma_i^2 - \sigma_v^2 \tag{8.19}$$

where the singular value, σ_{si}, characterizes the signal component.

Using the subspace concept [22] and SVD of the correlation matrix, $\tilde{\mathbf{R}}_{yy}$, the test statistics can approximate as

$$\hat{T}(\mathbf{y}_k) = \sum_{i=1}^{r} |\hat{U}_i \mathbf{y}_k|^2 \tag{8.20}$$

where $\hat{\mathbf{U}} = [\hat{U}_1, \ldots, \hat{U}_{(r)}]$ contains r orthogonal eigenvectors having the r largest eigenvalues of $\tilde{\mathbf{R}}_{yy}$, whereas the remaining $(M - r)$ eigenvectors correspond to the eigenvalues (power-spectral densities) of the noise. Thus, the decision rule becomes

$$\hat{T}(\mathbf{y}_k) = \begin{cases} > \gamma & \text{for } H_1 \\ \leq \gamma & \text{for } H_o \end{cases} \tag{8.21}$$

where the value of γ is very small and is set to be $\gamma = 0.025$ here. However, the accuracy of the method is less sensitive to the variation of γ.

Figure 8.1 illustrates the decision functions $T(\mathbf{y}_k)$ for different SNRs. It indicates that the decision function does not change with different SNRs to make the proposed method insensitive to the level of noise and the fixed threshold.

8.3.2 Selection of r

The r value is chosen based on EMD [23] decomposition of the input signal, $y(t)$, followed by the similarity measure between first IMF (Intrinsic Mode Function) g_1 and the jth IMF, g_j. The Euclidean distance between g_1 and g_j, d_{1j} is calculated to measure the similarity $s_{1j} = -d_{1j}^2$, and thus all similarities take negative values; the closer to zero the higher the similarity. Then, we calculate the preference, p which is set to some (25 %) percentile of the distribution of similarity among IMFs. Finally, the r value is chosen empirically as $r = 3$ if $p < 100$ or $r = 2$ if $100 \leq p \leq 200$ or $r = 1$ if $p > 200$.

8.3.3 Parameters Used

In our simulations, we have used the following parameters giving the best results: sparsity parameter, $\lambda = 0.4$, and rank of NMF decomposition, $Rank = 4$.

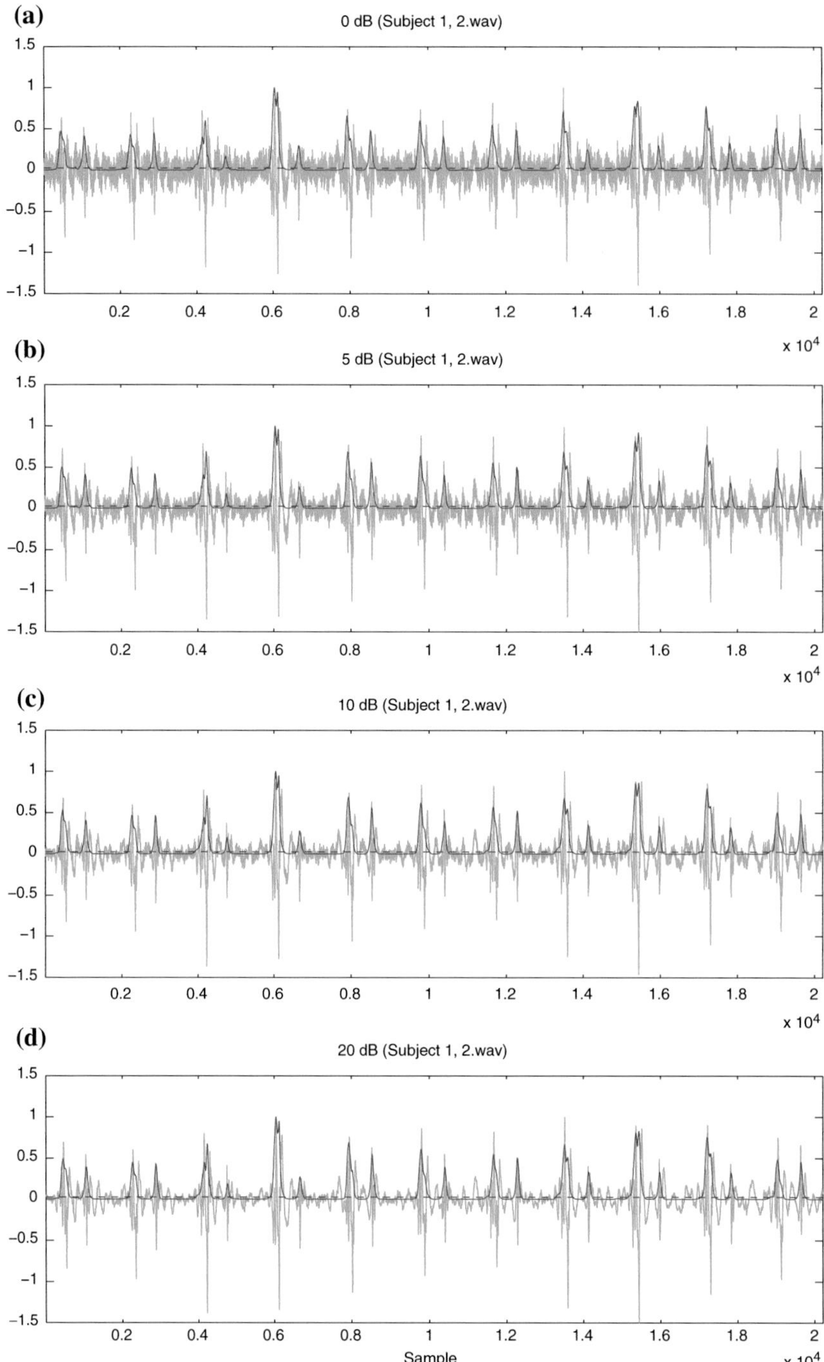

Fig. 8.1 Illustrative results of the decision functions for noisy PCG signals; **a** 0 dB, **b** 5 dB, **c** 10 dB, **d** 20 dB

8.4 Evaluation of the Proposed Method

8.4.1 Database

Two prototypes have dedicated to the MARS500 project promoted by ESA (European Space Agency). The prototype stethoscopes comprising an acoustic chamber in which a sound sensor is inserted. Electronics of signal conditioning and amplification are inserted in a case along with a Bluetooth standard communication module. The sounds are recorded with 16 bits accuracy and 8000 Hz sampling frequency in a wave format, using the software "Stetho" used under Alcatel-Lucent license. The dataset contains 50 sounds that correspond to the 6 Mars 500 volunteers. Each subject has 8–9 cardiac auscultations that have been collected periodically over a period of 520 days, as a part of the CardioPsy study (ESA-P.I: Pr P. AUBERT).

8.4.2 Preprocessing

The preprocessing of the input signal consists of decimation by factor 4 followed by filtering by a highpass filter with cut-off frequency of 30 Hz eliminating the noise collected by the prototype stethoscope. The filtered signal is then refiltered in reverse direction to eliminate the time delay and the resulting signal is normalized to a value of 1.

8.4.3 Illustrative Results

Illustrative plots are presented in Fig. 8.2 for a noise-free PCG recording, showing the results in consecutive steps of the proposed method. It could be observed that the proposed method gives reliable segmentation result and the NMF-based decision function derived is less sensitive to the level of threshold.

Figure 8.3 demonstrates how effectively segmentation is performed for noisy PCG signals using our proposed method by exploiting both the time and frequency domain features using NMF. According to Fig. 8.3, the cardiac events (i.e., S1 and S2 components) at low SNR can be detected successfully in the presence of high simulated noise. In comparison, the segmentation results without NMF are depicted in Fig. 8.4, where the results give more false positives (FPs) compared to the results of the proposed method in Fig. 8.3.

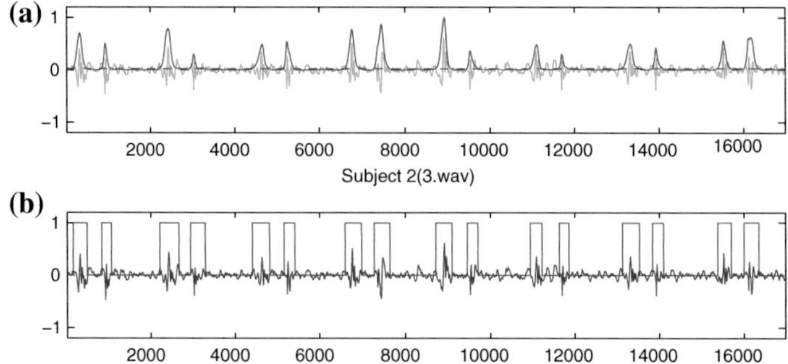

Fig. 8.2 An illustrative result of the proposed method for a noise-free PCG signal from subject 2 in the third recording: **a** Decision function with threshold and signal waveform; **b** segmentation results and signal waveform

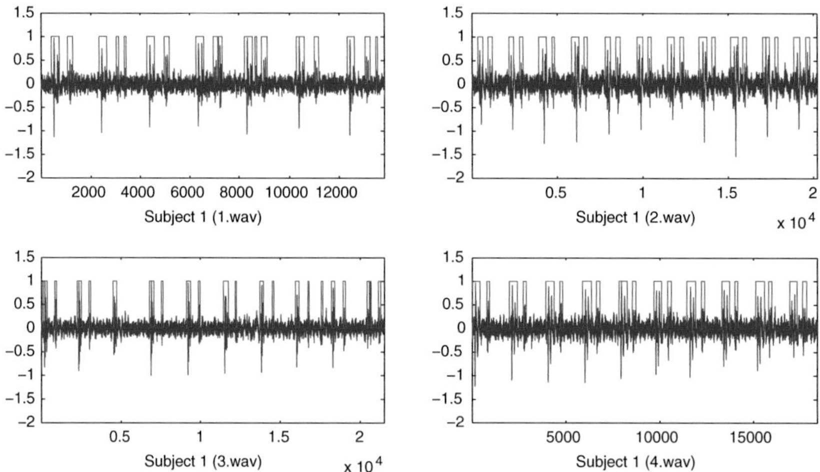

Fig. 8.3 Illustrative segmentation results for noisy PCG signals (0 dB SNR) by the proposed method

8.4.4 Performance Analysis

The performance has been measured as a method's ability to locate S1 and S2 correctly. It is performed by calculating the sensitivity (SE) and positive predictive value (PPV) as

$$\mathrm{SE} = \frac{\mathrm{TP}}{\mathrm{TP}+\mathrm{FN}} \tag{8.22}$$

$$\mathrm{PPV} = \frac{\mathrm{TP}}{\mathrm{TP}+\mathrm{FP}} \tag{8.23}$$

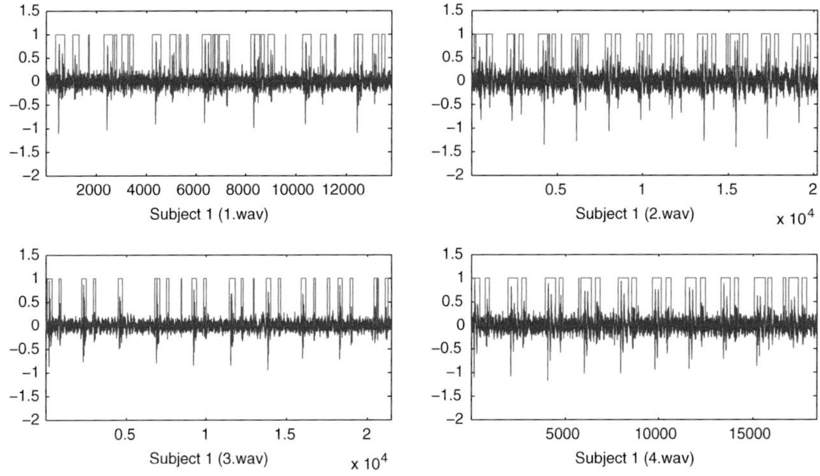

Fig. 8.4 Illustrative segmentation results for noisy PCG signals (0 dB SNR) without NMF

A sound is true positive (TP) or correctly located if the detected sound corresponds to a S1 or S2 sound predefined manually by the cardiologist, all other detected sounds are defined as false positive (FP) and all missed sounds are considered as false negative (FN) [24].

By implementing our proposed algorithms to the MARS500 database, a high overall sensitivity (SE) (96 %) and a positive predictive value (PPV) (93 %) are obtained based on the physician assessments (see Table 8.1 for detail results). According to the results in Table 8.1, the proposed method gives higher PPV for patients 4/5/6 which is noisier, while the results provide higher sensitivity for patients 1/2/3 which is cleaner. For comparison the performance of the method without NMF is presented

Table 8.1 Performance analysis of the proposed method

Subject	File index	TP	FN	FP	Sensitivity (%)	PPV (%)
1	1	14	0	2		
	2	22	0	0		
	3	19	1	1		
	4	20	0	0		
	5	16	0	0		
	6	30	0	8		
	7	22	0	0		
	8	27	0	4		
	9	18	0	3		
	Total	188	1	18	99.5	91.3

(continued)

Table 8.1 (continued)

Subject	File index	TP	FN	FP	Sensitivity (%)	PPV (%)
2	1	16	0	1		
	2	17	1	4		
	3	16	0	0		
	4	20	0	4		
	5	52	0	2		
	6	27	0	6		
	7	31	0	1		
	8	49	1	1		
	9	29	0	2		
	Total	257	2	21	99.2	92.4
3	1	30	4	1		
	2	25	0	2		
	3	29	0	2		
	5	34	0	1		
	6	30	1	2		
	7	31	0	11		
	8	37	0	7		
	9	23	0	7		
	Total	239	5	33	98.0	87.9
4	1	34	0	0		
	2	45	6	2		
	3	30	3	0		
	4	46	0	4		
	5	57	0	4		
	6	39	4	0		
	7	31	0	9		
	8	46	11	0		
	9	25	3	0		
	Total	353	27	19	92.9	94.9
5	1	15	15	0		
	2	27	1	1		
	3	23	0	3		
	5	28	9	0		
	6	25	1	0		
	7	22	0	4		
	8	34	0	3		
	9	21	0	3		
	Total	195	26	14	88.3	93.3

(continued)

Table 8.1 (continued)

Subject	File index	TP	FN	FP	Sensitivity (%)	PPV (%)
6	2	50	0	1		
	3	36	0	0		
	4	26	2	0		
	5	53	0	3		
	6	40	0	1		
	7	46	0	1		
	9	38	1	1		
	Total	289	3	7	98.9	97.6
Overall		1521	64	112	96.0	93.2

Table 8.2 Performance analysis without NMF

Subject	File index	TP	FN	FP	Sensitivity (%)	PPV (%)
1	1	14	0	6		
	2	22	1	2		
	3	19	1	3		
	4	20	0	1		
	5	17	0	3		
	6	23	7	12		
	7	22	0	12		
	8	27	7	6		
	9	18	6	8		
	Total	182	18	67	91.00	73.09
2	1	16	2	7		
	2	17	9	4		
	3	16	0	6		
	4	17	3	8		
	5	52	14	12		
	6	27	1	24		
	7	27	4	12		
	8	33	8	20		
	9	22	4	15		
	Total	227	45	108	83.46	67.76
3	1	28	5	14		
	2	25	0	13		
	3	19	8	14		
	5	30	3	10		
	6	26	4	9		
	7	20	10	7		
	8	28	9	17		
	9	18	10	8		
	Total	194	49	92	79.84	67.83

(continued)

Table 8.2 (continued)

Subject	File index	TP	FN	FP	Sensitivity (%)	PPV (%)
4	1	30	6	1		
	2	49	1	3		
	3	30	0	1		
	4	33	13	13		
	5	49	5	8		
	6	40	3	2		
	7	32	3	7		
	8	44	11	2		
	9	20	5	11		
	Total	297	41	47	87.87	86.34
5	1	26	7	6		
	2	25	2	13		
	3	21	2	11		
	5	32	6	1		
	6	25	1	2		
	7	24	0	9		
	8	26	2	32		
	9	21	0	14		
	Total	200	20	88	90.91	69.44
6	2	58	0	1		
	3	38	0	2		
	4	27	2	10		
	5	53	0	3		
	6	39	1	9		
	7	43	1	4		
	9	32	7	1		
	Total	290	11	30	96.35	90.63
Overall		1390	184	432	88.31	76.29

in Table 8.2, where the results show that the performance has been dropped significantly without NMF (SE: 88 %, PPV: 76 %).

8.5 Conclusion

Segmentation of narrowly spaced noisy PCG signals based on minimal a priori information, is presented. Here, the PCG segmentation is formulated as a detection problem and the start-point as well as end point of the S1 and S2 segments are determined using a proposed decision functions by time-scale transform and NMF-based likelihood ratio measures. In the proposed scheme, both the desired and extraneous signal-like components can be extracted from the high noise background. Moreover, the energy conservation of the Morlet wavelet in the proposed method keeps the average amplitude of the overall noisy PCG signal constant over both time and frequency domains [25]. The results based on real PCG data, are found promising for the proposed method. The advantages for the adoption of continuous wavelet transform and NMF together can be found under the high noisy condition and various subjects. Due to the promising performance, this NMF-based segmentation approach has potential medical applications for accurate detection of heart diseases based on the located S1 and S2 components.

Our future work aims to develop a subject independent system for the analysis of difficult PCG data with the aid of this new segmentation method proposed here. Moreover, we would like to extend this NMF-based scheme so that both denoising and PCG segmentation can be performed simultaneously within an integrated NMF-based framework by designing different analysis and synthesis dictionaries (basis functions) in the NMF scheme.

References

1. A. Moukadem, A. Dieterlen, N. Hueber, C. Brandt, Comparative study of heart sounds localization, bioelectronics, biomedical and bio-inspired systems. SPIE N 8068A-27, Prague (2011)
2. C.S. Lima, A. Tavares, J. Correia, M.J. Cardoso, D. Barbosa, in *New Developments in Biomedical Engineering*, ed. by Domenico Campolo, ISBN 978-953-7619-57-2 (InTech Publisher), p. 37–72 (2010)
3. H. Liang, S. Lukkarinen, I. Hartimo, *Heart Sound Segmentation Algorithm Based on Heart Sound Envelogram* (Helsinki University of Technology, Espoo, 1997)
4. C. Ahlström, Nonlinear Phonocardiographic Signal Processing. Institutionen för Medicinsk Teknik, 2008
5. H. Naseri, M.R. Homaeinezhad, H. Pourkhajeh, Noise/spike detection in phonocardiogram signal as a cyclic random process with non-stationary period interval. Comput. Biol. Med. **43**, 1205–1213 (2013)
6. S. Ari, K. Hembram, G. Saha, Detection of cardiac abnormality from PCG signal using LMS based least square SVM classifier. Expert Syst. Appl. **37**(12), 8019–8026 (2010)
7. H. Hadi, M. Mashor, M. Suboh, M. Mohamed, Classification of heart sound based on S-transform and neural networks, in *International Conference on Information Science, Signal Processing and Their Applications* (2010)
8. D.D. Lee, H.S. Seung, Algorithms for non-negative matrix factorization. Adv. Neural Inf. Proc. Syst. **13**, 556–562 (2001)
9. Y.-X. Wang, Y.-J. Zhang, Nonnegative matrix factorization: a comprehensive review. IEEE Trans. Knowl. Data Eng. **25**(6), 1336–1353 (2013)
10. A.K. Kattepur, F. Jin, F. Sattar, Single channel source separation for convolutive mixture with application to respiratory sounds, in *IEEE-EMBS Conference on Bio-inspired Systems and Signal Processing (BIOSIGNALS)* (2010)
11. B. Ghoraani, S. Krishnan, Time-frequency matrix feature extraction and classification of environmental audio signals. IEEE Trans. Audio Speech Lang. Process. **19**(7), 2197–2209 (2011)
12. M. Kim, J. Yoo, K. Kang, S. Choi, Nonnegative matrix partial co-factorization for spectral and temporal drum source separation. IEEE J. Sel. Top. Signal Process. **5**(6), 1192–1204 (2011)

13. G. Shah, P. Koch, C.B. Papadias, On the blind recovery of cardiac and respiratory sounds. IEEE J. Biomed. Health Inform. **19**(1), 151–157 (2015)
14. P.O. Hoyer, Non-negative sparse coding, in *IEEE Workshop Neural Networks for Signal Process* (2002), pp. 557–565
15. J. Eggert, E. Kmrner, Sparse coding and NMF, in *IEEE Conference on Neural Networks*, vol. 4 (2004), pp. 2529–2533
16. C.J. Lin, Projected gradient methods for non negative matrix factorization. Neural Comput. **19**(10), 2756–2779 (2007)
17. D. Kim, S. Sra, and I.S. Dhillon, Fast newton-type methods for the least squares non negative matrix approximation problem, in *SIAM Conference on Data Mining* (2007)
18. C. Torrence, G.P. Compo, A practical guide to wavelet analysis. Bull. Am. Meterol. Soc. **79**(1), 61–76 (1998)
19. L.L. Scharf, *Statistical Signal Processing: Detection Estimation and Time Series Analysis* (Addison-Wesley, Boston, 1991)
20. S.M. Kay, *Fundamentals of Statistical Signal Processing: Detection Theory* (Prentice Hall, Englewood Cliffs, 1998)
21. N. Lee, Q. Huynh, S. Schwartz, New methods of linear time-frequency analysis for signal detection, in *IEEE Proceedings on Time-Frequency and Time-Scale Analysis*, Paris, June 1996, pp. 13–16
22. A.-J. Van der Veen, E.D.F. Deprettere, A.L. Swindlehurst, Sub-space based signal analysis using singular value decomposition. Proc. IEEE **81**(9), 1277–1307 (1993)
23. B. Huang, A. Kunoth, An optimization based empirical mode decomposition scheme. J. Comput. Appl. Math. **240**(2013), 174–183 (2013)
24. A. Moukadem, A. Dieterlen, N. Hueber, C. Brandt, A robust heart sounds segmentation module based on S-transform. Biomed. Signal Process. Control **8**, 273–281 (2013)
25. S. Ventosa, C. Simon, M. Schimmel, J. Danobeitia, A. Manuel, The S-transform from a wavelet point of view. IEEE Trans. Signal Process. **56**(7), 2771–2780 (2008)

Printed in the United States
By Bookmasters